Gas Sensors

Cesare Malagù · Giulia Zonta

Gas Sensors

Principles, Physics, and Technology

Cesare Malagù
Department of Physics and Earth Sciences
University of Ferrara
Ferrara, Italy

Giulia Zonta
Department of Physics and Earth Sciences
University of Ferrara
Ferrara, Italy

ISBN 978-3-032-21613-7 ISBN 978-3-032-21614-4 (eBook)
https://doi.org/10.1007/978-3-032-21614-4

© The Editor(s) (if applicable) and The Author(s), under exclusive license to Springer Nature Switzerland AG 2026

This work is subject to copyright. All rights are solely and exclusively licensed by the Publisher, whether the whole or part of the material is concerned, specifically the rights of translation, reprinting, reuse of illustrations, recitation, broadcasting, reproduction on microfilms or in any other physical way, and transmission or information storage and retrieval, electronic adaptation, computer software, or by similar or dissimilar methodology now known or hereafter developed.
The use of general descriptive names, registered names, trademarks, service marks, etc. in this publication does not imply, even in the absence of a specific statement, that such names are exempt from the relevant protective laws and regulations and therefore free for general use.
The publisher, the authors and the editors are safe to assume that the advice and information in this book are believed to be true and accurate at the date of publication. Neither the publisher nor the authors or the editors give a warranty, expressed or implied, with respect to the material contained herein or for any errors or omissions that may have been made. The publisher remains neutral with regard to jurisdictional claims in published maps and institutional affiliations.

This Springer imprint is published by the registered company Springer Nature Switzerland AG
The registered company address is: Gewerbestrasse 11, 6330 Cham, Switzerland

If disposing of this product, please recycle the paper.

The most exciting phrase to hear in science, the one that heralds new discoveries, is not "Eureka!" but "That's funny"...

—Isaac Asimov

This book is dedicated to the memory of Celso Aldao and Giuliano Martinelli, friends and mentors who inspired the contents of this book.

Competing Interests The authors have no competing interests to declare that are relevant to the content of this manuscript.

Contents

About the Authors

Cesare Malagù was born in 1974. He has been coordinating the sensors group at the University of Ferrara since 2010 and is an associate professor since 2016. His research is based on modeling of transport phenomena in nanostructured semiconductors. His expertise regards thick-film gas sensors applied to environmental monitoring and cancer screening. His h-index is 40. He has more than 100 papers in peer-reviewed journals. He is in the editorial board of several journals.

Giulia Zonta was born in 1988. She is a Researcher in Physics at the Sensor Laboratory of the University of Ferrara. She works in the field of semiconductor physics, in the research and development on nanostructured chemoresistive gas sensors, particularly for medical applications. She is also a co-founder of SCENT S.r.l., a company born from university research focused on the identification of tumor biomarkers in human body fluids, aimed at developing non-invasive preventive screening devices based on gas sensor technology.

List of Figures

List of Tables

Chapter 1
Introduction

The study of solid-state physics and its applications in sensing technologies requires a rigorous foundation that bridges fundamental quantum mechanics with the physics of real materials. This textbook is designed to provide such a pathway, guiding the reader from the essential principles of quantum theory to the specific mechanisms underlying semiconductor devices and, in particular, metal oxide (MOX) gas sensors and their applications. By integrating rigorous theoretical formulations with physical interpretation, the book is intended to function both as a reference resource for researchers and as an effective didactic support for graduate-level education.

The first part of the volume establishes the necessary background in quantum mechanics and solid-state physics. Starting from operator formalism and symmetry considerations, the discussion progresses to crystal lattices, reciprocal space, and diffraction phenomena, culminating in Bloch's theorem and the band structure formalism. These foundations are essential for understanding how electrons behave in periodic potentials and for constructing the theoretical framework of semiconductors.

Building on this basis, the second part of the book addresses the electronic properties of semiconductors and their junctions. Topics such as Fermi energy, Fermi level, effective mass, band bending, and charge carrier statistics are presented to develop a comprehensive description of semiconductor behaviour. Special attention is devoted to the physics of p-n junctions and metal-semiconductor contacts, which represent the cornerstones of modern electronic and optoelectronic devices.

The third part focuses on chemoresistive gas sensors, highlighting their operation principles and the models employed to describe their electrical response to gaseous environments. Both classical and advanced approaches are considered, including Arrhenius-plot analyses, depletion approximations applied at the nanoscale, tunneling mechanisms, and the critical role of defects such as oxygen vacancies. Emphasis is placed on the phenomenon of in-diffusion and its consequences for sensing performance, as it represents a critical factor in determining the stability and functionality of MOX-based devices. Experimental techniques, such as scanning tunneling spectroscopy, are also introduced to connect theory with practice.

© The Author(s), under exclusive license to Springer Nature Switzerland AG 2026

C. Malagù and G. Zonta, *Gas Sensors*, https://doi.org/10.1007/978-3-032-21614-4_1

Finally, the last part of this book focuses on the structure of gas sensors, their fabrication methods, and the main characterization techniques. It also presents the most used semiconductor materials and the instrumental setups adopted in both laboratory and field applications, with examples of devices that integrate gas sensors for different practical purposes. In addition, temperature and humidity sensors are here discussed, which are essential for monitoring the environmental conditions in which gas sensors operate, since their response is strongly influenced by both parameters. An overview of the experimental setups for both laboratory-based and portable implementations is also provided.

Overall, the book offers a coherent journey from fundamental physics to applied chemoresistive gas sensor technology. It is intended not only to support the training of graduate students in advanced solid-state physics and materials science courses, but also to provide researchers with a structured overview of theoretical models and experimental methods relevant to semiconductor devices and gas sensing applications.

Chapter 2
Basics of Quantum Mechanics

> We call a real dynamical variable whose eigenstates form a
> complete set
> an observable (Paul Dirac).

This Chapter covers fundamental concepts in quantum mechanics, focusing on operators, symmetries, and their physical implications. In quantum systems, observables correspond to Hermitian operators, yielding real eigenvalues and orthogonal eigenvectors that form a complete basis. Measurement collapses a quantum state into an eigenstate of the operator involved within the standard Copenhagen formulation. The Chapter also discusses unitary operators, their role in transformations, and their relationship to Hamiltonian symmetry, with conserved quantities associated with these symmetries as per Noether's theorem, such as energy, linear momentum, and angular momentum. The linear momentum operator, linked to translational symmetry, has plane waves as eigenfunctions, yielding a parabolic dispersion relation for free particles with mass. Further, the commutation properties of operators provide conditions for simultaneous diagonalization, connecting to the uncertainty principle. The wave vector in k-space is introduced as essential for representing energy as a function of momentum, a central concept for analyzing the properties of solids. This Chapter is informed by the scientific literature cited below and reflects the authors' independent synthesis and interpretation of these sources [1–12].

2.1 Symmetries and Operators

In physics, an observable is any quantity that can be measured either directly (i.e., using appropriate measuring instruments) or indirectly (i.e., through calculations). In quantum mechanics, an observable is represented by a linear operator A, which must also be **Hermitian**:

$$A^{\dagger} = A$$

© The Author(s), under exclusive license to Springer Nature Switzerland AG 2026
C. Malagù and G. Zonta, *Gas Sensors*, https://doi.org/10.1007/978-3-032-21614-4_2

It has been demonstrated that a Hermitian operator has:

- *real eigenvalues*; in fact, if a' is the eigenvalue of A with eigenvector $|a'\rangle$ (normalized, i.e., $\langle a'|a'\rangle = 1$), the mean value of A is: $\langle a'|A|a'\rangle = a'\langle a'|a'\rangle = a'$; so, being A Hermitian, $\langle a'|A|a'\rangle = \langle a'|A|a'\rangle^* = a'^*$ and, by equating the two, it results that $a' = a'^*$, only valid when $a' \in \mathbb{R}$;
- *orthogonal eigenvectors* when they refer to diverse eigenvalues; in fact, by taking two different eigenvalues of A $(a' \neq a'')$ with eigenvectors $|a'\rangle$ and $|a''\rangle$, respectively: $\langle a'|A|a''\rangle = a''\langle a'|a''\rangle$ and, for hermiticity, $\langle a'|A|a''\rangle = \langle a''|A|a'\rangle^* = a'\langle a''|a'\rangle^* = a'\langle a'|a''\rangle$; by equating the two, it results that $(a'' - a')\langle a'|a''\rangle = 0$ and, being $a' \neq a''$, the only way for this equation to be null is that $\langle a'|a''\rangle = 0$ (**orthogonality condition**).

The eigenvectors of a Hermitian operator form a complete basis set. Therefore, to fully describe a physical system, including all its possible states, one must utilize a set of mutually compatible operators. Measurement corresponds to the application of an operator, causing the system's state to collapse into one of the operator's eigenstates, yielding a real eigenvalue as the measurement result (since the outcome of any measurement is a real number). An operator U can be defined as **unitary** if it satisfies the condition:

$$U^\dagger U = UU^\dagger = I,$$

where $U^\dagger$ denotes the Hermitian adjoint of U. A necessary and sufficient condition for an operator to be unitary, for continuous unitary transformations connected to the identity, is that it can be expressed in the exponential form:

$$U = e^{iA\varepsilon},$$

where A (the generator) is a Hermitian operator and ε a small real parameter. A acts as the generator of a linear continuous transformation, as the parameter ε can be made arbitrarily small, thereby ensuring the continuity of the transformation.

The transformation of a generic operator B under a unitary operation U is given by:

$$U^\dagger BU = B',$$

where B' denotes the **transformed** operator. If and only if $B' = B$, the transformation is also a symmetry of the operator U.

Of particular interest are the symmetries of the Hamiltonian, for which:

$$U^\dagger HU = H.$$

In classical physics, **Noether's theorem** states that every symmetry of the action corresponds to a conserved quantity. This connection can be readily understood in the quantum framework by multiplying both sides of the above equation by U:

$$UU^{\dagger}HU = U,$$

which simplifies to

$$HU = UH,$$

or equivalently,

$$[H, U] = 0.$$

Hence, H commutes with the operator U. Therefore, a symmetry of the Hamiltonian commutes with the Hamiltonian itself. Mathematically, if two operators commute, they can be simultaneously diagonalized, meaning that a common basis (and therefore a complete set of eigenvectors) can be chosen for both. From a physical standpoint, this concept is closely related to the Poisson bracket in classical mechanics. For a generic function f:

$$\frac{df}{dt} = [H, f] + \frac{\partial f}{\partial t}.$$

If the quantity is not explicitly time-dependent, the partial derivative term vanishes. The transition to quantum mechanics is straightforward: the classical function f is replaced by a generic operator A, and the Poisson bracket is replaced by the commutator, yielding:

$$\frac{dA}{dt} = \frac{[H, A]}{i\hbar} + \frac{\partial A}{\partial t}.$$

In the **Schrödinger representation**, one often considers "frozen" operators, whose form does not explicitly depend on time (differently from the **Heisenberg representation**), while the quantum states evolve. In this case, $\partial A/\partial t = 0$, and the total time derivative of such an operator reduces to:

$$\frac{dA}{dt} = \frac{[H, A]}{i\hbar}.$$

Thus, for operators that are not explicitly time-dependent, the total derivative is entirely determined by the commutator with the Hamiltonian. Consequently, if an operator A commutes with H, it represents a **constant of motion**.

If S is a symmetry of the Hamiltonian, such as $[S, H] = 0$, then S is conserved. Moreover, any such symmetry operator can be expressed as:

$$S = e^{ik\varepsilon},$$

where k is the Hermitian infinitesimal generator of the symmetry, and being ε potentially very small, S can be expanded in Taylor series as:

$$S = e^{ik\varepsilon} = \sum_{n=0}^{\infty} \frac{(ik\varepsilon)^n}{n!} = 1 + ik\varepsilon - \frac{k^2\varepsilon^2}{2} - i\frac{k^3\varepsilon^3}{6} + \frac{k^4\varepsilon^4}{24} \pm \ldots \sim 1 + ik\varepsilon.$$

By stopping the series at first order, it results that

$$[1 + ik\varepsilon, H] = 0$$

and

$$[k, H] = 0,$$

since the identity operator trivially commutes with H. Therefore, if S is a symmetry of the Hamiltonian, its infinitesimal generator k is conserved. Examples of this principle include:

- time homogeneity: conservation of total energy (Hamiltonian as generator of time evolution);
- space homogeneity: conservation of linear momentum (generator of translations);
- spatial isotropy: conservation of angular momentum (generator of rotations).

Finally, the generator of a symmetry can be simultaneously diagonalized with the Hamiltonian, reflecting the compatibility of conserved quantities with the system's dynamics.

2.2 Linear Momentum Operator

Consider the unitary operator $T(\varepsilon)$, referred to as the **infinitesimal translation operator**, which generates a translation of a wavefunction by an infinitesimal displacement ε. Its action on a wavefunction $\phi(x)$ is defined as:

$$T(\varepsilon)\phi(x) = \phi(x + \varepsilon).$$

On the other hand, the inverse translation corresponds to a negative displacement:

$$T(-\varepsilon) = T^{\dagger},$$

therefore:

$$T^{\dagger}T = I = TT^{\dagger}.$$

showing that T is unitary and that $T^{\dagger} = T^{-1}$. Expanding $\phi(x+\varepsilon)$ in a Taylor series and retaining only the first-order term yields:

$$\phi(x+\varepsilon) \approx \phi(x) + \varepsilon \frac{\partial \phi(x)}{\partial x}.$$

Comparing this expression with the operator definition of $T(\varepsilon)$, it follows that:

$$T(\varepsilon) = 1 + \varepsilon \frac{\partial}{\partial x} \equiv e^{\varepsilon \frac{\partial}{\partial x}},$$

because spatial translation operator is a symmetry, and thus it can also be written in exponential form as:

$$T(\varepsilon) = e^{ik\varepsilon},$$

where k must be Hermitian to guarantee unitarity. Since ε represents a spatial translation, classical mechanics dictates that its generator is the linear momentum. Thus, k encodes information about the linear momentum operator. Equating the arguments of the two exponential representations of $T(\varepsilon)$ gives:

$$\frac{\partial}{\partial x} = ik$$

and, equivalently,

$$k = -i\frac{\partial}{\partial x}.$$

Furthermore, multiplying both sides by $\hbar$, it results:

$$p_x = \hbar k = -i\hbar \frac{\partial}{\partial x},$$

which represents the linear momentum responsible for generating translations along the x-direction. The extension to three-dimensional space is straightforward. In the coordinate representation, the momentum operator takes the form:

$$\vec{p} = -i\hbar\vec{\nabla}.$$

As an illustrative case, consider a free particle described by the wavefunction $e^{i\vec{k}\cdot\vec{r}}$. We now verify whether this state corresponds to a well-defined momentum eigenstate. Applying the momentum operator $\vec{p}$ yields:

$$\vec{p}e^{i\vec{k}\cdot\vec{r}} = -i\hbar\vec{\nabla}e^{i\vec{k}\cdot\vec{r}} = -i\hbar i\vec{k}e^{i\vec{k}\cdot\vec{r}} = \hbar\vec{k}e^{i\vec{k}\cdot\vec{r}},$$

showing that $\hbar\vec{k}$ is the eigenvalue of the momentum operator. Hence, the eigenfunctions of $\vec{p}$ are plane waves. In this context, only the time-independent part of the Schrödinger equation (related to energy) has been considered, while the time-dependent part is governed by the time evolution of the Hamiltonian. For a free particle, the total energy is purely kinetic and given by $p^2/2m$. Applying the kinetic-energy operator to the momentum eigenstate, and noting that $p^2 = -\hbar^2\nabla^2$, we obtain:

$$\frac{p^2}{2m}e^{i\vec{k}\cdot\vec{r}} = \frac{-\hbar^2\nabla^2}{2m}e^{i\vec{k}\cdot\vec{r}} = \frac{-\hbar^2}{2m}(-k^2)e^{i\vec{k}\cdot\vec{r}} = \frac{\hbar^2\mathrm{k}^2}{2m}e^{i\vec{k}\cdot\vec{r}}.$$

The fact that $\left[\vec{p}, p^2\right] = 0$ implies that, being $e^{i\vec{k}\cdot\vec{r}}$ eigenfunction of $\vec{p}$, it must also be eigenfunction of $p^2/2m$ (i.e., the Hamiltonian of a free particle). Therefore, it is possible to diagonalize the free particle system using a set of simultaneous eigenfunctions of $\vec{p}$ and H, which are precisely plane waves. Consequently, any eigenfunction can be expressed as a superposition of plane waves. These plane wave eigenstates yield the following eigenvalues for the momentum and the Hamiltonian:

$$\vec{p} = \hbar\vec{k}$$

and

$$E = \frac{\hbar^2\mathrm{k}^2}{2m},$$

respectively. The latter relation constitutes the **dispersion relation** of a free particle of mass m. Unlike the linear dispersion relation found in classical electromagnetism (where energy is proportional to momentum), the dispersion for a free particle with a mass m is parabolic, as illustrated in Fig. 2.1.

The two operators can be simultaneously specified with arbitrary precision, since no uncertainty principle applies between them. At every continuous point along

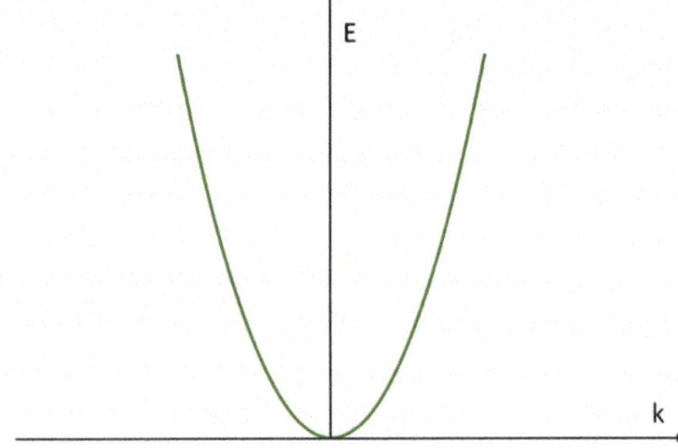

Fig. 2.1 Dispersion relation in one dimension for a free particle with a mass m. The ordinate represents the energy $E = \hbar^2k^2/2m$, whereas the abscissa corresponds to the wave-vector k

the curve (parabola), it is therefore possible to measure both the momentum and the energy simultaneously. The vector $\vec{k}$ in $e^{i\vec{k}\cdot\vec{r}}$, which acts as the generator of spatial translations, is referred to as the **wave vector**. In summary, d'Alembert's equation admits plane-wave solutions with a linear dispersion relation, as it describes massless particles. In contrast, here we consider free particles with mass; in this case, the Hamiltonian reduces to kinetic energy, which commutes with the momentum operator. As a result, the eigenfunctions remain plane waves, but they now correspond to massive particles and lead to a quadratic dispersion relation. As an example, let us apply the commutator $[p_x, x]$ to an operator A:

$$[p_x, x]A = i\hbar\frac{\partial}{\partial x}xA - xi\hbar\frac{\partial A}{\partial x} = xi\hbar\frac{\partial A}{\partial x} + i\hbar A - xi\hbar\frac{\partial A}{\partial x} = i\hbar A,$$

which differs from zero, while:

$$[p_y, x]A = 0.$$

Thus:

$$[p_i, x_j] = i\hbar\delta_{ij}.$$

Measuring momentum and position in the same direction results in a commutator that is non-zero. By deriving this from Schwarz's inequality applied to Hilbert space, one can establish the uncertainty principle within this matrix formalism. The crucial point is that when the commutator of two operators is non-zero, simultaneous diagonalization is impossible, leading to inherent indeterminacy expressed as:

$$\Delta p_x \Delta x \geq \frac{\hbar}{2}.$$

Consequently, if one attempts to construct an analogous dispersion relation with momentum on the ordinate axis and position on the abscissa, a continuous function cannot be defined (i.e., there is no continuous representation of momentum as a function of position). For this reason, all band-structure representations discussed in this course are based on the dispersion relation $\varepsilon_n(\vec{k})$, expressing energy as a function of momentum, rather than of position x. The k-space thus contains the essential physical information regarding the solid.

References

1. Ashcroft NW, Mermin ND (1976) Solid state physics. Holt, Rinehart and Winston, New York
2. Sakurai JJ, Napolitano J (2017) Modern quantum mechanics, 2nd edn. Cambridge University Press, Cambridge

3. Dirac PAM (1958) The principles of quantum mechanics, 4th edn. Oxford University Press, Oxford
4. Cohen-Tannoudji C, Diu B, Laloë F (1977) Quantum mechanics, vols I–II. Wiley, New York
5. Landau LD, Lifshitz EM (1977) Quantum mechanics: non-relativistic theory, 3rd edn. Pergamon Press, Oxford
6. Shankar R (1994) Principles of quantum mechanics, 2nd edn. Springer, New York
7. Greiner W, Müller B (1994) Quantum mechanics: symmetries. Springer, Heidelberg
8. Kittel C (2005) Introduction to solid state physics, 8th edn. Wiley, Hoboken
9. Goldstein H, Poole C, Safko J (2002) Classical mechanics, 3rd edn. Addison-Wesley, San Francisco
10. Noether E (1918) Invariante Variationsprobleme. Nachrichten von der Gesellschaft der Wissenschaften zu Göttingen, Mathematisch-Physikalische Klasse 1918:235–257
11. Wigner EP (1939) On unitary representations of the inhomogeneous Lorentz group. Ann Math 40:149–204
12. Stone M (2000) The physics of quantum fields. Springer, New York, NY

Chapter 3
Introduction to Solid State Physics

> Geometry is the archetype of the beauty of the world. (Johannes Kepler)

For a solid to qualify as a crystal, two components are required: a Bravais lattice and an atomic basis (concepts that will be detailed in this Chapter). Different elements can exhibit identical Bravais lattice structures but may possess unique bases. The atomic basis, in turn, influences the element's valence, which determines the number of electrons available for conduction. Thus, the conductive or insulating nature of a material depends not only on the crystal's symmetry but also on its atomic basis. In this Chapter, foundational concepts of solid-state physics are introduced, providing essential knowledge for understanding the physics underlying semiconductor gas sensors. The content of this Chapter results from an integrated and critical elaboration of the literature, mainly drawing on the references listed below, together with original analyses and interpretations developed by the authors [1–9].

3.1 Bravais Lattice

A **lattice** is defined as an array of points in space that forms a periodic structure, representing a mathematical abstraction, as schematized in Fig. 3.1.

A lattice is defined as a **Bravais lattice** if its points are arranged such that the system appears symmetric from any lattice point. When positioned at any point within the lattice, one observes the same structural arrangement (relative positions of other points) regardless of the specific point chosen, reflecting its discrete periodicity. Mathematically, the position of each point in a 3D-space can be specified by a vector $\vec{R}$:

$$\vec{R} = n_1\vec{a}_1 + n_2\vec{a}_2 + n_3\vec{a}_3,$$

where $n_i \in Z$. The vectors $\vec{a}_i$ (with $i = 1, 2, 3$) represent a basis of the 3D-space and they do not lie on a single plane, although they do not necessarily form an orthonormal

© The Author(s), under exclusive license to Springer Nature Switzerland AG 2026
C. Malagù and G. Zonta, *Gas Sensors*, https://doi.org/10.1007/978-3-032-21614-4_3

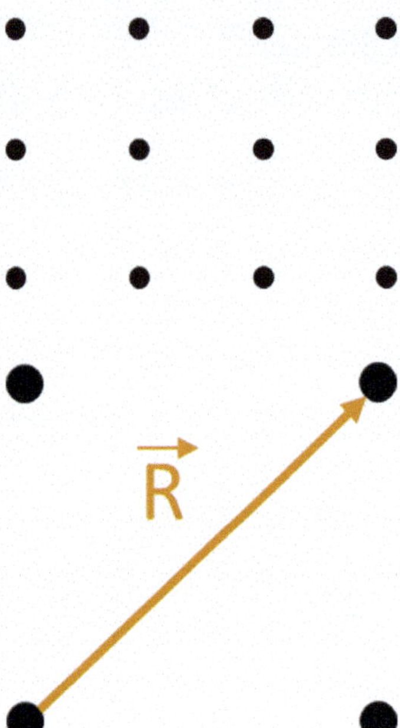

Fig. 3.1 Lattice structure schematization as a periodic set of points in space

Fig. 3.2 Vector $\vec{R}$ connecting two lattice points

basis. Any point $\vec{R}$ within this space can be expressed as a linear combination of these basis vectors. By defining one point as the origin, all other points can be reached through $\vec{R}$. An example is shown in Fig. 3.2.

In this context, we are considering a discrete symmetry, meaning that all translations result exclusively in lattice points, with no translations ending at non-lattice points, as shown in Fig. 3.3.

To classify a lattice, it is essential to introduce the concept of a **primitive cell**. A primitive cell is defined as the smallest volume element which, when translated throughout space, fills the entire crystal without overlapping. By construction, each primitive cell contains exactly one lattice point within its volume. The vectors $\vec{a}_i$ are named "primitive" vectors when they generate the cell of minimum volume. Denoting by n the lattice point density and by v the primitive cell volume, one has:

$$n = \frac{N}{V}$$

and

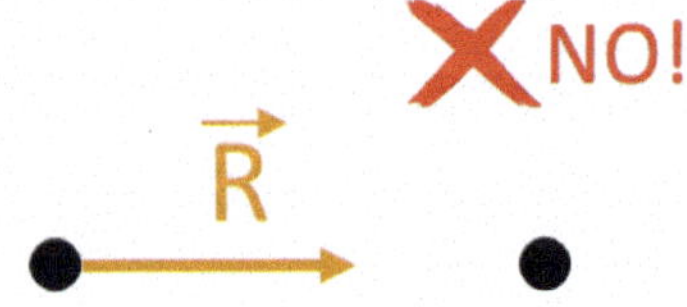

Fig. 3.3 Translations whose endpoints do not coincide with lattice points are not considered allowable lattice translations

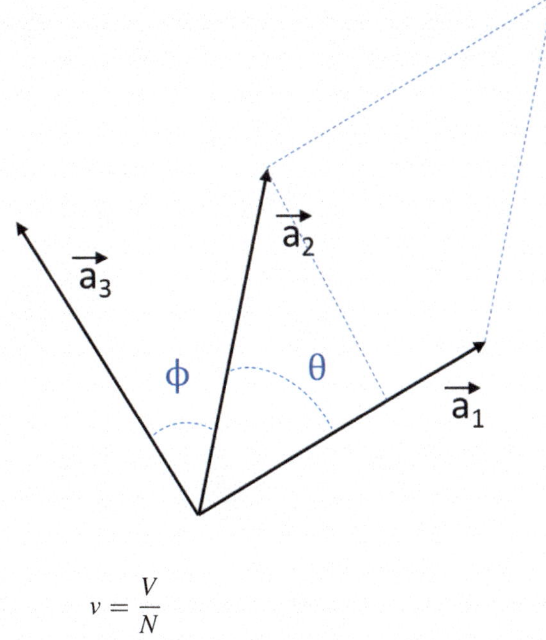

Fig. 3.4 Primitive vectors $\vec{a}_i$ in three-dimensional space

$$v = \frac{V}{N}$$

respectively, where N is the total number of cells in a solid of volume V. It follows directly that $nv = 1$, or equivalently, $v = 1/n$. In a three-dimensional representation, the primitive vectors $\vec{a}_i$ emanate from a common origin and are linearly independent, i.e., they do not lie on the same plane. The cross product $\vec{a}_1 \times \vec{a}_2$ yields the area of the parallelogram spanned by these two vectors (see Fig. 3.4). The scalar triple product:

$$\vec{a}_3 \cdot \vec{a}_1 \times \vec{a}_2 = v$$

defines the volume of the primitive cell, where $\vec{a}_3$ acts as the height relative to the base.

For a finite crystal, the maximum number of cells along each of the three primitive directions is given by the integers N_1, N_2, and N_3, whose product is typically of the order of Avogadro's number (approximately $10^7 - 10^8$ for each dimension) and therefore extremely large. This justifies treating real crystals as infinite Bravais lattices for practical purposes. The total number of lattice points in the crystal is then:

$$N = N_1 N_2 N_3.$$

Since $v = V/N = 1/n$ and n depends solely on the geometry of the lattice, it remains constant.

Besides the primitive cell, one may also define a **unit** or **conventional cell**, when overlaps are allowed. The unit cell is not necessarily primitive and may contain more than one lattice point. A unit cell is chosen to exhibit the full symmetry of the crystal, although its volume exceeds that of the primitive cell. For example, in

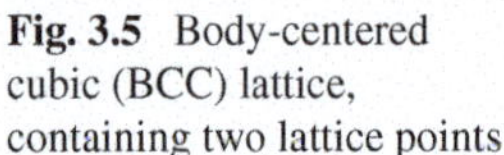

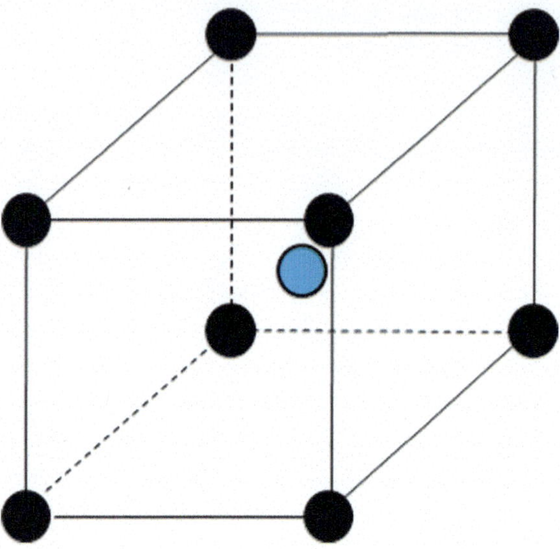

Fig. 3.5 Body-centered cubic (BCC) lattice, containing two lattice points

a body-centered cubic (BCC) lattice (see Fig. 3.5), a lattice point is located at each cube corner and one at the cube center. Although the representation might suggest nine lattice points per cell, the actual number is:

$$\frac{1}{8}(8) + 1 = 2,$$

because each corner point is shared among eight adjacent cubes. Hence, the BCC unit cell contains two lattice points, whereas the primitive cell contains only one. Consequently, the unit cell volume is twice that of the primitive cell.

Another important cell type is the **Wigner–Seitz cell**, which combines the features of both a primitive and a unit cell. It fills space without overlaps, preserves all the symmetries of the lattice, and contains exactly one lattice point, with a volume $v = 1/n$.

3.2 Reciprocal Lattice

In this Section, the formulation of the system within the framework of the **reciprocal lattice** is introduced. In physics, the reciprocal lattice is defined as the Fourier transform of the direct lattice, and it provides a representation of spatial periodicity in real space. To construct the reciprocal lattice, a basis must be specified in momentum space, derived from the real-space primitive vectors. The reciprocal lattice vectors $\vec{b}_i$ are defined as functions of the real-space primitive vectors $\vec{a}_i$:

$$\vec{b}_1 = 2\pi \frac{\vec{a}_2 \times \vec{a}_3}{\vec{a}_1 \cdot \vec{a}_2 \times \vec{a}_3},$$

$$\vec{b}_2 = 2\pi \frac{\vec{a}_3 \times \vec{a}_1}{\vec{a}_1 \cdot \vec{a}_2 \times \vec{a}_3},$$

$$\vec{b}_3 = 2\pi \frac{\vec{a}_1 \times \vec{a}_2}{\vec{a}_1 \cdot \vec{a}_2 \times \vec{a}_3}.$$

If the vectors $\vec{a}_1, \vec{a}_2, \vec{a}_3$ are linearly independent and non-coplanar, the same holds for the reciprocal vectors $\vec{b}_1$, $\vec{b}_2$, $\vec{b}_3$. The two sets of vectors are connected by the important following relation:

$$\vec{a}_i \cdot \vec{b}_j = 2\pi \delta_{ij}.$$

For instance, $\vec{a}_2 \cdot \vec{b}_1 = \vec{a}_3 \cdot \vec{b}_1 = 0$, while $\vec{a}_1 \cdot \vec{b}_1 = 2\pi$. Analogous relations hold for $\vec{b}_2$ and $\vec{b}_3$.

To establish the connection between the direct and reciprocal lattices, consider a plane wave (associated with a crystal electron) of the form $e^{i\vec{k}\cdot\vec{r}}$. Imposing the symmetry under translations of a lattice vector $\vec{R}$, valid for $\forall \vec{R}$:

$$e^{i\vec{k}\cdot\vec{r}} = e^{i\vec{k}\cdot(\vec{r}+\vec{R})},$$

this leads to the condition:

$$e^{i\vec{k}\,\vec{R}} = 1.$$

For $\vec{k}$ to be a reciprocal lattice vector, it must satisfy the condition:

$$\vec{k} \cdot \vec{R} = 2\pi n,$$

with $n \in Z$. The vector $\vec{k}$ can therefore be expressed as a linear combination of the reciprocal basis vectors:

$$\vec{k} = k_1\vec{b}_1 + k_2\vec{b}_2 + k_3\vec{b}_3,$$

with $k_i \in Z$, leading to:

$$\vec{k} \cdot \vec{R} = k_1 n_1 + k_2 n_2 + k_3 n_2 = N,$$

with $N \in Z$. Thus, the reciprocal lattice is itself a Bravais lattice, spanned by non-coplanar basis vectors with integer coefficients.

Next, we examine the relationship between the volume v of the primitive cell in real space and the volume $\vec{b}_1 \cdot (\vec{b}_2 \times \vec{b}_3)$ in $\vec{k}$-space. Using the vector identity:

$$\vec{A} \times \vec{B} \cdot \vec{C} \times \vec{D} = \vec{D} \cdot \left(\vec{A} \times \vec{B} \times \vec{C}\right) = \left[\left(\vec{A} \cdot \vec{C}\right)\vec{B} - \left(\vec{B} \cdot \vec{C}\right)\vec{A}\right] \cdot \vec{D}$$

and applying it to the $\vec{k}$-space volume:

$$\begin{aligned}
\vec{b}_1 \cdot \vec{b}_2 \times \vec{b}_3 &= 2\pi \frac{\vec{a}_2 \times \vec{a}_3}{\vec{a}_1 \cdot \vec{a}_2 \times \vec{a}_3} \cdot \vec{b}_2 \times \vec{b}_3 \\
&= \frac{2\pi}{\vec{a}_1 \cdot \vec{a}_2 \times \vec{a}_3} \vec{b}_3 \cdot \left(\vec{a}_2 \times \vec{a}_3 \times \vec{b}_2\right) \\
&= \frac{2\pi}{\vec{a}_1 \cdot \vec{a}_2 \times \vec{a}_3} \vec{b}_3 \cdot \left[\left(\vec{a}_2 \cdot \vec{b}_2\right)\vec{a}_3 - \left(\vec{a}_3 \cdot \vec{b}_2\right)\vec{a}_2\right] \\
&= \frac{2\pi}{\vec{a}_1 \cdot \vec{a}_2 \times \vec{a}_3} \vec{b}_3 \cdot \left(\vec{a}_2 \cdot \vec{b}_2\right)\vec{a}_3,
\end{aligned}$$

we end up with the following relation:

$$\vec{b}_1 \cdot \vec{b}_2 \times \vec{b}_3 = \frac{(2\pi)^3}{\vec{a}_1 \cdot \vec{a}_2 \times \vec{a}_3}.$$

This quantity represents the volume of the primitive cell of the reciprocal lattice. It also corresponds to the volume of the Wigner–Seitz cell in reciprocal space, which preserves the crystal symmetry and is known as the **first Brillouin zone**. Importantly, a small primitive cell in real space implies a large Brillouin zone volume, and vice versa. This reciprocal relationship links the direct space (defined by vectors of dimension $[L]$) to momentum space, where vectors have dimension $\left[L^{-1}\right]$.

3.3 Bragg Diffraction

In crystalline solids, intense peaks of scattered radiation, referred to as **Bragg peaks**, are observed at specific wavelengths and incident angles. A crystal, together with its lattice, can be decomposed into sets of lattice planes arranged in different orientations, each characterized by a constant interplanar spacing d, as shown in Fig. 3.6.

Consider radiation of wavelength λ incident on two successive crystal planes, as illustrated in Fig. 3.7. For a pronounced peak in the intensity of the scattered radiation, two conditions must be simultaneously satisfied:

1. **specular reflection** at the lattice planes, such that the angle of incidence equals the angle of reflection with respect to the plane normal;
2. **constructive interference** of the reflected rays originating from successive planes at the detection screen.

The path difference between the two reflected rays (labeled 1 and 2 in Fig. 3.7) corresponds to the projection of ray 1 onto the direction of ray 2 and is equal to $2d\sin\theta$. Constructive interference (i.e., the condition under which the two rays reach

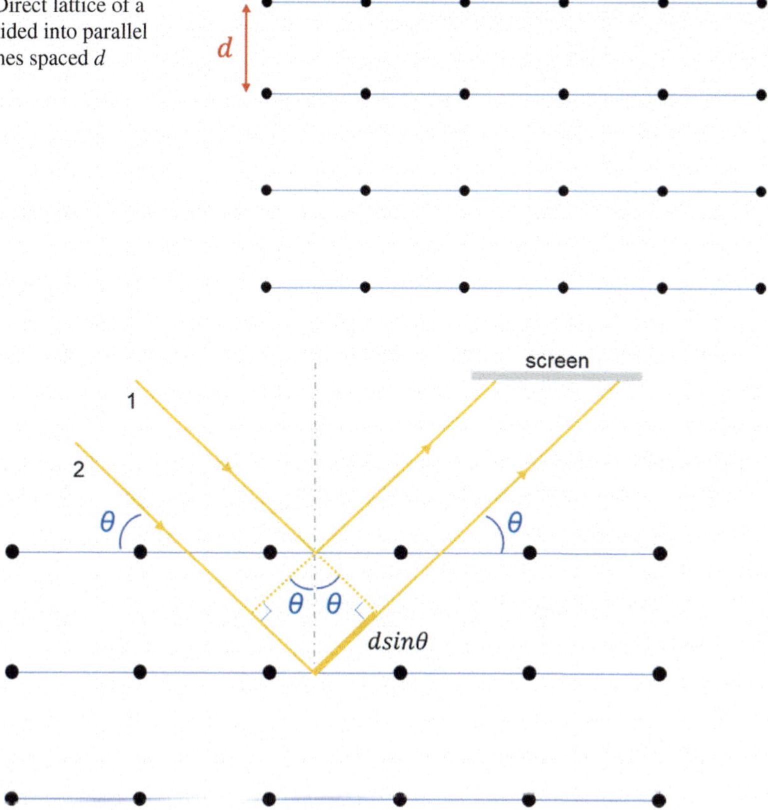

Fig. 3.6 Direct lattice of a crystal divided into parallel lattice planes spaced d

Fig. 3.7 Bragg reflection from two adjacent lattice planes

the detector in phase) requires that this path difference equals an integer multiple of the wavelength:

$$2d \sin \theta = n\lambda,$$

where n denotes the order of the corresponding reflection. There are multiple methods to partition the crystal into planes, each resulting in additional reflections. This relationship is commonly referred to as the **Bragg equation**. An alternative but equivalent formulation of diffraction in crystals is provided by the von Laue approach, which will be addressed in the next Section.

3.4 Von Laue Formulation

The **Von Laue formulation** offers an alternative perspective on the Bragg treatment by operating within the framework of reciprocal lattice space. There are two main distinctions from the Bragg approach:

1. there is no requirement to partition the crystal into specific lattice planes;
2. there is no assumption of specular reflection imposed.

In this context, the crystal is viewed as consisting of identical microscopic entities (e.g., ions, atoms) located at the sites $\vec{R}$ of a Bravais lattice, each capable of scattering the incident radiation in all directions. Sharp peaks in the scattered intensity are observed only in those directions and at those wavelengths where the rays scattered from all lattice points interfere constructively. To illustrate this, let us first consider two scatterers. The incident wavevector is denoted as:

$$\vec{k} = \frac{2\pi}{\lambda}\hat{n},$$

while the scattered wavevector is represented as:

$$\vec{k}' = \frac{2\pi}{\lambda}\hat{n}'.$$

As illustrated in Fig. 3.8, the path difference between the two rays (1 and 2), determined by the projection of the second ray onto the first, can be expressed as:

$$d\cos\theta + d\cos\theta' = \vec{d}\cdot\hat{n} - \vec{d}\cdot\hat{n}',$$

where $\vec{d}$ is the distance between the two points considered, and θ, θ' are the angles between the direction of the incident and scattered rays and the vector $\vec{d}$. For constructive interference to occur, this path difference must equal an integer multiple m of the wavelength:

$$\vec{d}\cdot\left(\hat{n} - \hat{n}'\right) = m\lambda.$$

Expressing the wavevectors in terms of the wavelength, this becomes:

$$\vec{d}\cdot\left(\vec{k} - \vec{k}'\right) = 2\pi m.$$

Now, extending this relationship to an array of scatterers located at the sites of a Bravais lattice (involving more than two points), the above condition must be satisfied simultaneously for all values of $\vec{d}$, that correspond to Bravais lattice vectors. Therefore, we replace $\vec{d}$ with $\vec{R}$ in the previous equation, yielding:

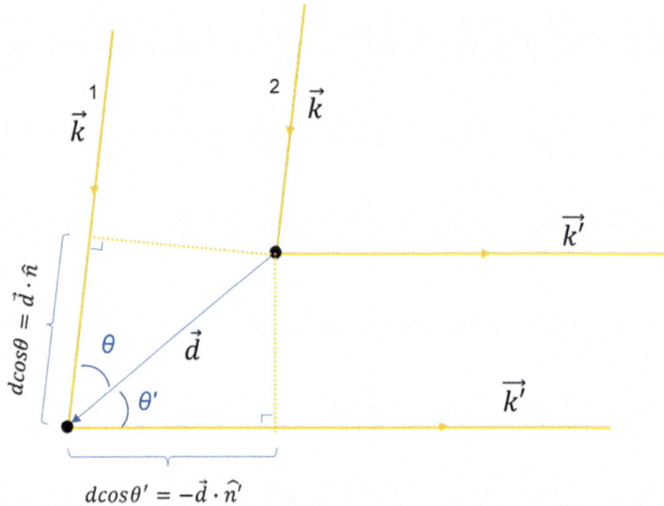

Fig. 3.8 Von Laue diffraction for two scatterers separated by a distance $\vec{d}$. The incident and scattered wavevectors are $\vec{k} = (2\pi/\lambda)\hat{n}$ and $\vec{k}' = (2\pi/\lambda)\hat{n}'$; θ and θ' are the angles between the rays and the vector $\vec{d}$

$$\vec{R}\cdot\left(\vec{k}-\vec{k}'\right) = 2\pi m$$

or equivalently,

$$e^{i\left(\vec{k'}-\vec{k}\right)\cdot\vec{R}} = 1.$$

Given that for a generic reciprocal lattice vector $\vec{G}$, the identity $e^{i\vec{G}\cdot\vec{R}} = 1$ holds, it follows that:

$$\vec{G} = \vec{k}' - \vec{k}$$

must be a reciprocal lattice vector (and likewise $\vec{k} - \vec{k}'$). Additionally, as for elastic scattering, $\vec{k}$ and $\vec{k}'$ have the same magnitude k:

$$k = \left|\vec{k}'\right| = \left|\vec{k} - \vec{G}\right|.$$

Squaring both sides of this equation gives:

$$k^2 = k^2 + G^2 - 2\vec{k}\cdot\vec{G},$$

and from this relationship, we can derive the Von Laue condition as follows:

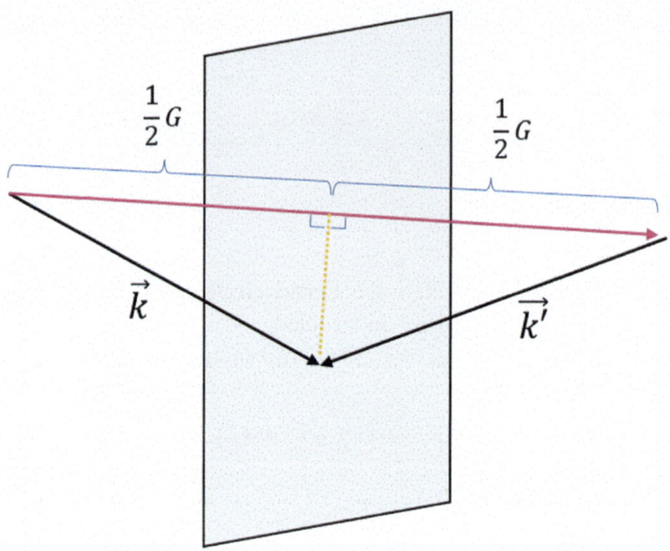

Fig. 3.9 Schematic representation of the Von Laue formulation: the projection of $\vec{k}$ along the direction $\hat{G}$ has a magnitude equal to $(1/2)G$, with the tip of the vector located on a Bragg plane that bisects $\vec{G}$

$$\vec{k} \cdot \hat{G} = \frac{1}{2}G.$$

As shown in Fig. 3.9, this indicates that the component of the incident wavevector $\vec{k}$ along the reciprocal lattice vector $\vec{G}$ must be half the length of $\vec{G}$. An incident vector satisfies the von Laue condition if and only if the tip of the vector lies in a plane that serves as the perpendicular bisector of the line connecting the origin of $\vec{k}$-space to a reciprocal lattice point $\vec{k}'$. These planes are referred to as **Bragg Planes**.

It can be demonstrated that that Bragg and Von Laue formulations are equivalent. Specifically, the $\vec{k}$-space Bragg plane associated with a particular diffraction peak in the Von Laue formulation corresponds to the family of direct lattice Bragg planes responsible for the peak in the Bragg formulation. Reconsidering the specular reflection of a ray on a direct lattice (as shown in Fig. 3.10), it is possible to construct the vector $\vec{G}$ in direct space, which is perpendicular to the Bragg plane of the direct crystal. The equivalence between the two formulations arises from the relationship between the vectors of the reciprocal lattice and the family of direct lattice planes. Assuming that $\vec{k}$ and $\vec{k}'$ have the same magnitude (i.e., $\lambda = \lambda'$) and satisfy the von Laue condition with $\vec{G} = \vec{k}' - \vec{k}$, both vectors make the same angle θ with the plane perpendicular to the vector $\vec{G}$. Consequently, the scattering can be interpreted as a Bragg reflection at an angle θ from the family of direct lattice planes that are perpendicular to the reciprocal lattice vector $\vec{G}$.

Mathematically, $\vec{G}$ is an integer multiple of the shortest reciprocal lattice vector $\vec{G}_0$ and is perpendicular to it, such that:

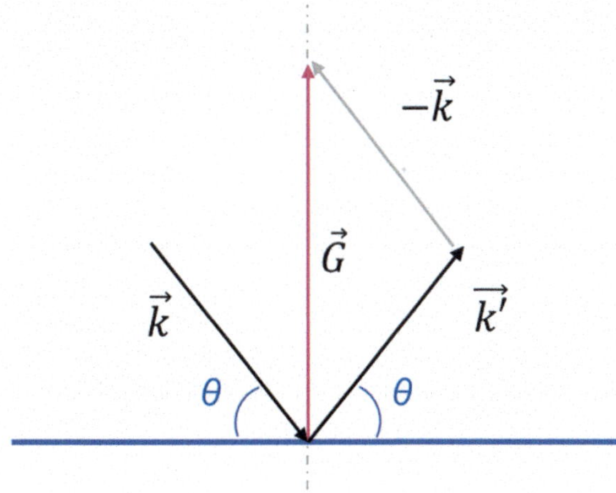

Fig. 3.10 Reflection of a ray with wavevector $\vec{k}$ from a Bragg plane of a direct lattice

$$G_0 = \frac{2\pi}{d}$$

and

$$G = \frac{2\pi}{d} n,$$

where n is an integer. From Fig. 3.10, it results that $G = 2k \sin\theta$, which leads to the equation:

$$\frac{2\pi}{d} n = 2\left(\frac{2\pi}{\lambda}\right) \sin\theta.$$

Simplifying, this yields:

$$n\lambda = 2d \sin\theta,$$

that is the Bragg law. Thus, a Laue diffraction peak corresponding to a change in wavevector described by the reciprocal lattice vector $\vec{G}$, is associated with a Bragg reflection from the family of direct lattice planes that are perpendicular to $\vec{G}$. The order n of Bragg reflection is given by:

$$n = \frac{G}{G_0}.$$

3.5 Bloch's Theorem

Let us now consider an extension of the dispersion relation previously established for free particles, focusing on the behaviour of electrons (massive particles) within a lattice. We have already introduced all the necessary concepts, including those related to waves and electromagnetism. The understanding of wave phenomena has been instrumental in describing free quantum particles, whose eigenstates are plane waves obeying a quadratic dispersion relation. We now turn to the case of a particle that is not completely free but instead moves within an infinite lattice characterized by discrete translational symmetries. Our goal is to examine how to reformulate the Hamiltonian in this context in a straightforward manner. Starting from the free-particle Hamiltonian, which consists solely of the kinetic energy term, we introduce a potential $V(\vec{r})$ that depends on position:

$$H = \frac{p^2}{2m} + V(\vec{r}).$$

The potential must exhibit periodicity in real space, satisfying the condition $V(\vec{r}) = V\left(\vec{r} + \vec{R}\right)$, which reflects the discrete symmetry of the lattice. This is the fundamental condition imposed by the periodic structure. In analogy with the treatment of free particles, we can define a translation operator T, which now depends on the discrete lattice vectors $\vec{R}$ rather than on a continuous parameter, since continuous translations would break the lattice symmetry. Thus, we define the operator as:

$$T = T\left(\vec{R}\right).$$

When applied to a generic position-dependent function $f(\vec{r})$, by definition:

$$T\left(\vec{R}\right)f(\vec{r}) \equiv f\left(\vec{r} + \vec{R}\right).$$

This operation yields a new function $f\left(\vec{r} + \vec{R}\right)$. Furthermore, applying both $T\left(\vec{R}\right)$ and $T\left(-\vec{R}\right)$ to the same function gives:

$$T\left(\vec{R}\right)T\left(-\vec{R}\right)f(\vec{r}) = f\left(\vec{r} + \vec{R} - \vec{R}\right) = f(\vec{r})$$

so that the function remains unchanged. Hence, the inverse and the Hermitian conjugate of the operator T can be defined (it is a unitary operator):

$$T\left(-\vec{R}\right) = T^{\dagger}\left(\vec{R}\right)$$

which implies:

$$T\left(-\vec{R}\right)T\left(\vec{R}\right) = I = T\left(\vec{R}\right)T\left(-\vec{R}\right).$$

Now, let us consider the Schrödinger equation:

$$H(\vec{r})\varphi(r) = E\varphi(r)$$

and apply the translation operator T directly to it:

$$T\left(\vec{R}\right)H(\vec{r})\varphi(\vec{r}) = H\left(\vec{r}+\vec{R}\right)\varphi\left(\vec{r}+\vec{R}\right).$$

Recalling that the Hamiltonian consists of a kinetic term (which remains invariant under translations) and a periodic potential, we see that the Hamiltonian itself is periodic:

$$H\left(\vec{r}+\vec{R}\right) = H(\vec{r}).$$

Hence, the application of T yields:

$$T\left(\vec{R}\right)H(\vec{r})\varphi(\vec{r}) = H(\vec{r})T\left(\vec{R}\right)\varphi(\vec{r}),$$

which implies:

$$T\left(\vec{R}\right)H(\vec{r}) = H(\vec{r})T\left(\vec{R}\right),$$

or equivalently:

$$TH - HT = 0$$

$$[T, H] = 0.$$

The translation operator T therefore commutes with the Hamiltonian H. This means that if we have a set of eigenfunctions of H, denoted by $\phi_{n\vec{k}}(\vec{r})$, this same set also constitutes a set of eigenfunctions of T. We can thus write:

$$\begin{cases} H\phi_{n\vec{k}}(\vec{r}) = \varepsilon_{n\vec{k}}\phi_{n\vec{k}}(\vec{r}) \\ T\left(\vec{R}\right)\phi_{n\vec{k}}(\vec{r}) = C\left(\vec{R}\right)\phi_{n\vec{k}}(\vec{r}). \end{cases}$$

Therefore, both the Hamiltonian and the finite periodic translation operator can be simultaneously diagonalized using the same set of eigenfunctions. Considering two lattice vectors $\vec{R}_1$ and $\vec{R}_2$, we can apply the additivity property of the operator:

$$T\left(\vec{R}_1\right)T\left(\vec{R}_2\right)\phi_{n\vec{k}}(\vec{r}) = \phi_{n\vec{k}}\left(\vec{r}+\vec{R}_1+\vec{R}_2\right) = T\left(\vec{R}_1+\vec{R}_2\right)\phi_{n\vec{k}}(\vec{r}).$$

Applying this relation to the eigenvalues gives:

$$C\left(\vec{R}_1\right)C\left(\vec{R}_2\right) = C\left(\vec{R}_1+\vec{R}_2\right).$$

Extending this reasoning to n successive translations, we find:

$$\left[C\left(\vec{R}\right)\right]^n = C\left(n\vec{R}\right).$$

Considering now a general lattice vector $\vec{R} = n_1\vec{a}_1 + n_2\vec{a}_2 + n_3\vec{a}_3$, we can express the general coefficient representing the eigenvalue of the operator as a complex number:

$$C(\vec{a}_i) = e^{2\pi i x_i}$$

where x_i can be a complex number. Any complex number can be represented in this exponential form; if x_i is real, the coefficient represents a pure phase factor (but this is not a priori known, so a complex number can always be written as the exponential of another complex number). Hence, in the general case we can express:

$$C\left(\vec{R}\right) = e^{2\pi i n_i x_i}.$$

Now, considering a generic vector in momentum space (not restricted to the reciprocal lattice), it can always be expressed as a linear combination of the basis vectors of reciprocal space, although not necessarily with integer coefficients:

$$\vec{k} = x_1\vec{b}_1 + x_2\vec{b}_2 + x_3\vec{b}_3.$$

Taking the scalar product gives:

$$\vec{k}\cdot\vec{R} = 2\pi n_i x_i,$$

where summation is implied. So, the coefficient can be written as:

$$C\left(\vec{R}\right) = e^{i\vec{k}\cdot\vec{R}}.$$

Thus, the eigenvalue can be represented as a function of both reciprocal and direct space vectors. Applying the operator T to the eigenfunction, we obtain:

$$T\left(\vec{R}\right)\phi_{n\vec{k}}(\vec{r}) = C\left(\vec{R}\right)\phi_{n\vec{k}}(\vec{r}) = e^{i\vec{k}\cdot\vec{R}}\phi_{n\vec{k}}(\vec{r}),$$

which leads directly to:

$$\phi_{n\vec{k}}\left(\vec{r}+\vec{R}\right)=e^{i\vec{k}\cdot\vec{R}}\phi_{n\vec{k}}(\vec{r}).$$

This is the fundamental formulation of **Bloch's theorem**. The wavefunctions can equivalently be expressed as plane waves modulated by a position-dependent function:

$$\phi_{n\vec{k}}(\vec{r})=e^{i\vec{k}\cdot\vec{r}}f(\vec{r})$$

which must satisfy the condition derived above. Considering translation by a lattice vector $\vec{R}$:

$$\phi_{n\vec{k}}\left(\vec{r}+\vec{R}\right)=e^{i\vec{k}\cdot\vec{R}}e^{i\vec{k}\cdot\vec{r}}f\left(\vec{r}+\vec{R}\right)=e^{i\vec{k}\cdot\vec{R}}e^{i\vec{k}\cdot\vec{r}}f(\vec{r}).$$

For this to hold, the function $f(\vec{r})$ must satisfy:

$$f\left(\vec{r}+\vec{R}\right)=f(\vec{r}).$$

So, for a set of eigenfunctions to take the form of a plane wave multiplied by a position-dependent function, the latter must be necessarily periodic. This leads to the second formulation of Bloch's theorem:

$$\begin{cases}\phi_{n\vec{k}}(\vec{r})=e^{i\vec{k}\cdot\vec{r}}u_{n\vec{k}}(\vec{r})\\ u_{n\vec{k}}(\vec{r})=u_{n\vec{k}}\left(\vec{r}+\vec{R}\right)\end{cases}$$

which again implies:

$$\phi_{n\vec{k}}\left(\vec{r}+\vec{R}\right)=e^{i\vec{k}\cdot\vec{R}}\phi_{n\vec{k}}(\vec{r}),$$

demonstrating the perfect equivalence between the two formulations of Bloch's theorem.

References

1. Ashcroft NW, Mermin ND (1976) Solid state physics. Holt, Rinehart and Winston, New York
2. Kittel C (2005) Introduction to solid state physics, 8th edn. Wiley, Hoboken
3. Ziman JM (1972) Principles of the theory of solids, 2nd edn. Cambridge University Press, Cambridge
4. Born M, Huang K (1996) Dynamical theory of crystal lattices. New York, NY; online edn, Oxford Academic, 31 Oct 2023
5. Bloch F (1929) Über die Quantenmechanik der Elektronen in Kristallgittern. Z Phys 52:555–600

6. Laue M (1952) Eine quantitative Prüfung der Theorie für die Interferenz-Erscheinungen bei Röntgenstrahlen. Naturwissenschaften 39:368–372
7. Bragg WL, Bragg WH (1913) The reflection of X-rays by crystals. Proc R Soc A 88:428–438
8. Harrison WA (1980) Solid state theory. Dover Publications, New York
9. Cohen-Tannoudji C, Diu B, Laloë F (1977) Quantum mechanics, vols I–II. Wiley, New York

Chapter 4
Electrons in a Weak Periodic Potential

To see a world in a grain of sand. (William Blake)

The approach described herein is applicable to nearly free electron metals, specifically those belonging to groups I, II, III, and IV of the periodic table. These metals possess an electronic structure characterized by *s* and *p* electrons located outside of a closed shell noble gas configuration. The discussion starts with an examination of the Sommerfeld free electron gas, which is subsequently adjusted to incorporate the effects of a weak periodic potential. Conduction bands in solids can indeed be considered as "free-electron-like" due to a couple of key reasons, which relate to the behaviour of conduction electrons in a crystalline lattice. The two reasons are:

1. the **Pauli Exclusion Principle**, which states that two fermions (like electrons) cannot occupy the same quantum states simultaneously. In the context of a solid, the orthogonality to the core states (ultimately enforced by the Pauli principle) increases the kinetic energy near the ion cores, effectively pushing conduction electrons into interstitial regions. Instead, the conduction electrons occupy states in the conduction band, which are further away from the core ions. This positioning allows them to behave as though they are "freer" because they are allowed to have higher energy states significantly distanced from the influence of the atomic cores;
2. the **Screening effect**, active in the presence of conduction electrons that effectively shield or "screen" the interaction between other conduction electrons and the positive ions of the lattice. This happens because the presence of multiple conduction electrons in a material leads to an averaged effect where the total positive potential from the ions is diminished in the region where conduction electrons are present. As a result, the conduction electrons move as though they are in a weak, slowly varying effective periodic potential. The effective potential experienced by individual conduction electrons is less than what it would be in a non-metallic material, contributing to their increased mobility, akin to free electrons.

© The Author(s), under exclusive license to Springer Nature Switzerland AG 2026

C. Malagù and G. Zonta, *Gas Sensors*, https://doi.org/10.1007/978-3-032-21614-4_4

Overall, these factors create a scenario where conduction electrons can be treated similarly to free electrons, allowing for the conduction of electricity within the material. This behaviour is fundamentally important for understanding electrical conduction in metals and semiconductors. The material discussed in this Chapter arises from a critical integration of previously published works, as cited below, together with original insights developed by the authors [1–14].

4.1 General Approach: Free Electron Case

When the periodic potential is zero, the solutions to Schrödinger equation are plane waves. Upon introducing a weak periodic potential, the starting point for the mathematical treatment is the expansion of the exact solution as a linear combination of plane waves. Let us consider the wavefunction of a Bloch state with crystal momentum $\vec{k}$:

$$\phi_{\vec{k}}(\vec{r}) = \sum_{\vec{K}} c_{\vec{k}-\vec{K}} e^{i\left(\vec{k}-\vec{K}\right)\cdot\vec{r}},$$

where, the coefficients $c_{\vec{k}-\vec{K}}$ and the corresponding energy eigenvalue ε are determined by the following set of equations:

$$\left[\frac{\hbar^2}{2m}\left(\vec{k}-\vec{K}\right)^2 - \varepsilon\right] c_{\vec{k}-\vec{K}'} + \sum_{\vec{K}'} U_{\vec{K}'-\vec{K}} c_{\vec{k}-\vec{K}'} = 0.$$

For a fixed vector $\vec{k}$, there exists one such equation for each reciprocal lattice vector $\vec{K}$. The (infinite) different solutions corresponding to a given $\vec{k}$ are conventionally labelled by the **band index** n.

In the case free electron case, where the periodic potential vanishes ($U_{\vec{K}} = 0$), the equation reduces to:

$$(\varepsilon^0_{\vec{k}-\vec{K}} - \varepsilon) c_{\vec{k}-\vec{K}} = 0,$$

with

$$\varepsilon^0_{\vec{q}} = \frac{\hbar^2}{2m}\vec{q}^{\,2}.$$

This is achieved by introducing the new the variable:

$$\vec{q} = \vec{k} - \vec{K}.$$

For each reciprocal lattice vector $\vec{K}$, the solution can be both

$$c_{\vec{k}-\vec{K}} = 0$$

or

$$\varepsilon = \varepsilon^0_{\vec{k}-\vec{K}}.$$

The solution can be also classified into two cases: non-degenerate or degenerate.

- In the **non-degenerate** case, a single vector $\vec{K}$ corresponds to a unique energy value $\varepsilon = \varepsilon^0_{\vec{k}-\vec{K}}$, with the corresponding wavefunction is a plane wave of the form:

$$\phi_{\vec{k}} \propto e^{i\left(\vec{k}-\vec{K}\right)\cdot\vec{r}}.$$

- In the **degenerate** case, there exist a set of m distinct reciprocal lattice vectors $\vec{K}_i$ $(i = 1, \ldots, m)$ satisfying:

$$\varepsilon^0_{\vec{k}-\vec{K}_1} = \ldots = \varepsilon^0_{\vec{k}-\vec{K}_m}.$$

When this condition holds, the corresponding eigenfunctions are m independent plane waves, and any linear combination of them is also a valid solution. Thus, there is complete freedom in choosing the coefficients $c_{\vec{k}-\vec{K}}$ for $\vec{K} = \vec{K}_1, \ldots, \vec{K}_m$.

Having analyzed the free-electron case, we now turn to the case where the periodic potential is nonzero, $U_{\vec{K}} \neq 0$. The introduction of such a potential couples plane waves whose wavevectors differ by a reciprocal lattice vector, leading to the splitting of degenerate energy levels and the formation of energy bands and band gaps, the essential features of the electronic structure in crystalline solids.

4.2 First Case: Non-Degenerate

Let us now consider the case in which the Fourier components of the periodic potential $U_{\vec{K}}$ are nonzero but very small, approaching zero in magnitude. For a fixed value of $\vec{k}$, consider a reciprocal lattice vector $\vec{K}_1$ such that the corresponding free-electron energy $\varepsilon^0_{\vec{k}-\vec{K}_1}$ is well separated from all other $\varepsilon^0_{\vec{k}-\vec{K}}$ values, by an amount large compared to the characteristic potential strength U. This condition can be expressed as:

$$\left|\varepsilon^0_{\vec{k}-\vec{K}_1} - \varepsilon^0_{\vec{k}-\vec{K}}\right| \gg U$$

for a fixed $\vec{k}$ and for all $\vec{K} \neq \vec{K}_1$ (see Fig. 4.1).

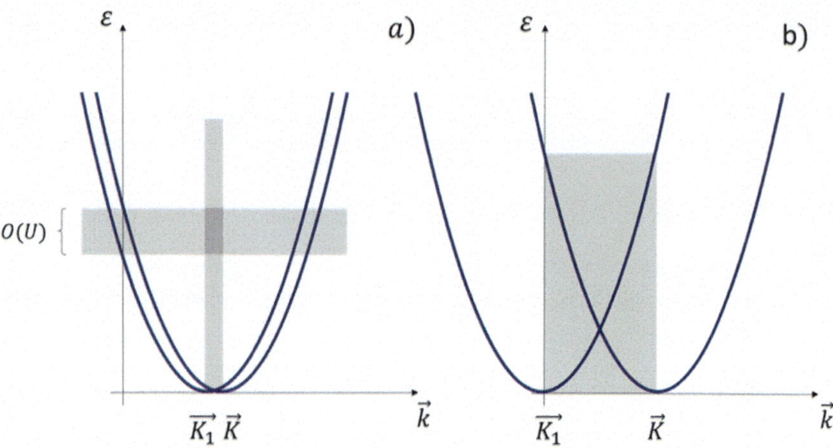

Fig. 4.1 (**a**) Case where the non-degenerate condition does not hold: $\left|\varepsilon^0_{\vec{k}-\vec{K}_1} - \varepsilon^0_{\vec{k}-\vec{K}}\right| \approx U$. (**b**) Case where the non-degenerate condition holds: $\left|\varepsilon^0_{\vec{k}-\vec{K}_1} - \varepsilon^0_{\vec{k}-\vec{K}}\right| \gg U$. Here, U refers to a typical Fourier component of the periodic potential

Let us now examine the effect of this weak potential on the free-electron energy level $\varepsilon = \varepsilon^0_{\vec{k}-\vec{K}_1}$, for which $c_{\vec{k}-\vec{K}} = 0$ when $\vec{K} \neq \vec{K}_1$. Setting $\vec{K} = \vec{K}_1$ in the general equation:

$$\left[\frac{\hbar^2}{2m}\left(\vec{k} - \vec{K}\right)^2 - \varepsilon\right]c_{\vec{k}-\vec{K}} + \sum_{\vec{K}'} U_{\vec{K}'-\vec{K}}\, c_{\vec{k}-\vec{K}'} = 0,$$

we obtain:

$$\left(\varepsilon - \varepsilon^0_{\vec{k}-\vec{K}_1}\right)c_{\vec{k}-\vec{K}_1} = \sum_{\vec{K}} U_{\vec{k}-\vec{K}_1} c_{\vec{k}-\vec{K}}.$$

Since the potential energy can be defined up to an arbitrary additive constant, we can set this constant so that the spatial average of the potential U_0 over a primitive cell vanishes, i.e., $U_{\vec{K}} = 0$ for $\vec{K} = 0$. Consequently, only terms with $\vec{K} \neq \vec{K}_1$ contribute to the right-hand side. We now consider the solution for which $c_{\vec{k}-\vec{K}}$ vanishes for $\vec{K} \neq \vec{K}_1$ as U approaches zero. In this case, the right-hand side of the above equation contains only terms of second order in U. To demonstrate this, let us rewrite the general equation for $\vec{K} \neq \vec{K}_1$, making the dependence on the coefficients explicit:

$$c_{\vec{k}-\vec{K}} = \frac{U_{\vec{K}_1-\vec{K}} c_{\vec{k}-\vec{K}_1}}{\varepsilon - \varepsilon^0_{\vec{k}-\vec{K}}} + \sum_{\vec{K}'\neq\vec{K}_1} \frac{U_{\vec{K}'-\vec{K}} c_{\vec{k}-\vec{K}'}}{\varepsilon - \varepsilon^0_{\vec{k}-\vec{K}}}$$

Among these terms, the one involving $\vec{K}_1$ is dominant, being of lower order in U than the rest. The assumption of non-degeneracy implies that $\varepsilon^0_{\vec{k}-\vec{K}_1}$ is not nearly degenerate to any other $\varepsilon^0_{\vec{k}-\vec{K}}$, so that the denominators $\varepsilon - \varepsilon^0_{\vec{k}-\vec{K}}$ are not of the order of U. Thus, the potential in the numerator does not cancel, and there are no other terms of comparable magnitude to the $\vec{K} = \vec{K}_1$ term. Consequently, we can approximate:

$$c_{\vec{k}-\vec{K}} = \frac{U_{\vec{K}_1-\vec{K}} c_{\vec{k}-\vec{K}_1}}{\varepsilon - \varepsilon^0_{\vec{k}-\vec{K}}} + O(U^2).$$

Substituting this expression in the general equation yields:

$$\left(\varepsilon - \varepsilon^0_{\vec{k}-\vec{K}_1}\right) c_{\vec{k}-\vec{K}_1} = \sum_{\vec{K}} U_{\vec{k}-\vec{K}_1} c_{\vec{k}-\vec{K}}$$

$$\left(\varepsilon - \varepsilon^0_{\vec{k}-\vec{K}_1}\right) c_{\vec{k}-\vec{K}_1} = \sum_{\vec{K}} \frac{U_{\vec{K}-\vec{K}_1} U_{\vec{K}_1-\vec{K}}}{\varepsilon - \varepsilon^0_{\vec{k}-\vec{K}}} c_{\vec{k}-\vec{K}_1} + O(U^3).$$

To find the perturbed energy, we set $\varepsilon = \varepsilon^0_{\vec{k}-\vec{K}_1}$ in the denominators. Recall that, for the free-electron case, $(\varepsilon^0_{\vec{k}-\vec{K}} - \varepsilon) c_{\vec{k}-\vec{K}} = 0$, so the energy shift induced by the potential is of second order in U. The resulting expression for the perturbed energy is:

$$\varepsilon = \varepsilon^0_{\vec{k}-\vec{K}_1} + \sum_{\vec{K}} \frac{\left|U_{\vec{K}-\vec{K}_1}\right|^2}{\varepsilon^0_{\vec{k}-\vec{K}_1} - \varepsilon^0_{\vec{k}-\vec{K}}} + O(U^3).$$

This second-order correction reflects the general tendency of energy levels to shift away from nearby states, although true band repulsion becomes relevant only in the presence of near degeneracy. Specifically, if $\varepsilon^0_{\vec{k}-\vec{K}} < \varepsilon^0_{\vec{k}-\vec{K}_1}$, the energy ε increases; whereas if $\varepsilon^0_{\vec{k}-\vec{K}} > \varepsilon^0_{\vec{k}-\vec{K}_1}$, the energy ε decreases. Hence, in the absence of near degeneracy, the energy shift induced by a weak periodic potential is a **second-order correction in** U.

4.3 Second Case: Nearly Degenerate

Consider m distinct lattice vectors:

$$\vec{K}_1, \vec{K}_2, \ldots, \vec{K}_m$$

such that the corresponding free-electron energies are:

$$\varepsilon^0_{\vec{k}-\vec{K}_1}, \varepsilon^0_{\vec{k}-\vec{K}_2,} \ldots, \varepsilon^0_{\vec{k}-\vec{K}_m},$$

all close to one another (within an energy range of order U) but well separated from all other energy values $\varepsilon^0_{\vec{k}-\vec{K}}$. This condition can be written as:

$$\left| \varepsilon^0_{\vec{k}-\vec{K}} - \varepsilon^0_{\vec{k}-\vec{K}_i} \right| \gg U$$

for $i = 1, \ldots, m$ and $\vec{K} \neq \vec{K}_1, \ldots, \vec{K}_m$.

In this situation the Schrödinger equation does not reduce to a single relation but instead produces a set of m coupled equations of the form:

$$\left[\frac{\hbar^2}{2m} \left(\vec{k} - \vec{K} \right)^2 - \varepsilon \right] c_{\vec{k}-\vec{K}} + \sum_{\vec{K}'} U_{\vec{K}'-\vec{K}} c_{\vec{k}-\vec{K}'} = 0$$

Separating the contributions associated with the nearly degenerate vectors $\vec{K}_1, \ldots, \vec{K}_m$ from those involving the remaining coefficients yields:

$$\left(\varepsilon - \varepsilon^0_{\vec{k}-\vec{K}_i} \right) c_{\vec{k}-\vec{K}_i} = \sum_{j=1}^{m} U_{\vec{K}_j-\vec{K}} c_{\vec{k}-\vec{K}_j} + \sum_{\vec{K}\neq\vec{K}_1,\ldots,\vec{K}_m} U_{\vec{K}-\vec{K}_i} c_{\vec{k}-\vec{K}}$$

with $i, j = 1, \ldots, m$.

In the limit of vanishing potential, the coefficients in the first term $c_{\vec{k}-\vec{K}_1}, \ldots, c_{\vec{k}-\vec{K}_m}$ remain finite, while all other coefficients $c_{\vec{k}-\vec{K}}$ ($\vec{K} \neq \vec{K}_1, \ldots, \vec{K}_m$) in the second term are of order U. Solving for these small coefficients gives:

$$c_{\vec{k}-\vec{K}} = \frac{1}{(\varepsilon - \varepsilon^0_{\vec{k}-\vec{K}})} \left(\sum_{j=1}^{m} U_{\vec{K}_j-\vec{K}} c_{\vec{k}-\vec{K}_j} + \sum_{\vec{K}'\neq\vec{K}_1,\ldots,\vec{K}_m} U_{\vec{K}'-\vec{K}} c_{\vec{k}-\vec{K}'} \right),$$

with $\vec{K} \neq \vec{K}_1, \ldots, \vec{K}_m$.

The structure of the equations corresponds to that of the coefficients $c_{\vec{k}-\vec{K}}$ in the case of no near degeneracy (see Sect. 4.2). Since these coefficients $c_{\vec{k}-\vec{K}}$ are of order U, the second sum in brackets contributes terms of order U^2. Hence, to the leading order we may write:

$$c_{\vec{k}-\vec{K}} = \frac{1}{(\varepsilon - \varepsilon^0_{\vec{k}-\vec{K}})} \sum_{j=1}^{m} U_{\vec{K}_j-\vec{K}} c_{\vec{k}-\vec{K}_j} + O(U^2).$$

Substituting this expression back into the full system gives:

$$\left(\varepsilon - \varepsilon^0_{\vec{k}-\vec{K}_i}\right) c_{\vec{k}-\vec{K}_i}$$
$$= \sum_{j=1}^{m} U_{\vec{K}_j-\vec{K}} c_{\vec{k}-\vec{K}_j} + \sum_{j=1}^{m} \left(\sum_{\vec{K} \neq \vec{K}_1,\ldots,\vec{K}_m} \frac{U_{\vec{K}-\vec{K}_i} U_{\vec{K}_j-\vec{K}}}{\varepsilon - \varepsilon^0_{\vec{k}-\vec{K}}} \right) c_{\vec{K}-\vec{K}_j} + O(U^3).$$

This yields a system of m coupled equations for the m unknown coefficients $c_{\vec{k}-\vec{K}_i}$, where the second term is of higher order in U than the first term. When there is no near degeneracy (i.e., $m = 1$), the system reduces to the single non-degenerate equation discussed previously in Sect. 4.2. To find the leading correction in U, we can simplify the equations by maintaining only the first term, as:

$$\left(\varepsilon - \varepsilon^0_{\vec{k}-\vec{K}_i}\right) c_{\vec{k}-\vec{K}_i} = \sum_{j=1}^{m} U_{\vec{K}_j-\vec{K}} c_{\vec{k}-\vec{K}_j}$$

where $i = 1, \ldots m$. These equations describe the coupling of m nearly degenerate quantum levels by periodic potential and determine the structure of the energy bands in regions where degeneracies occur (e.g., near the boundaries of the Brillouin zone).

4.4 Energy Level Near a Single Bragg Plane

We now focus on the case in which two electron energy levels lie within an energy interval of order U, while remaining well separated from all other free-electron energies by amounts much larger than U. Starting from the general system of equations describing m nearly degenerate states, we restrict ourselves to the special case $m = 2$ (i.e., near degeneracy of order two). The resulting dynamics is governed by the pair of coupled equations:

$$\begin{cases} \left(\varepsilon - \varepsilon^0_{\vec{k}-\vec{K}_1}\right) c_{\vec{k}-\vec{K}_1} = U_{\vec{K}_2-\vec{K}_1} c_{\vec{k}-\vec{K}_2} \\ \left(\varepsilon - \varepsilon^0_{\vec{k}-\vec{K}_2}\right) c_{\vec{k}-\vec{K}_2} = U_{\vec{K}_1-\vec{K}_2} c_{\vec{k}-\vec{K}_1}. \end{cases}$$

Introducing the variables:

$$\vec{q} = \vec{k} - \vec{K}_1$$

and

$$\vec{K} = \vec{K}_2 - \vec{K}_1,$$

the system becomes:

$$\begin{cases} \left(\varepsilon - \varepsilon^0_{\vec{q}}\right) c_{\vec{q}} = U_{\vec{K}} c_{\vec{q}-\vec{K}} \\ \left(\varepsilon - \varepsilon^0_{\vec{q}-\vec{K}}\right) c_{\vec{q}-\vec{K}} = U_{-\vec{K}} c_{\vec{q}} = U^*_{\vec{K}} c_{\vec{q}}. \end{cases}$$

Here the two free-electron energies satisfy the conditions

$$\varepsilon^0_{\vec{q}} \approx \varepsilon^0_{\vec{q}-\vec{K}}$$

and

$$\left|\varepsilon^0_{\vec{q}} - \varepsilon^0_{\vec{q}-\vec{K}'}\right| \gg U.$$

For all $\vec{K}' \neq \vec{K}, 0$. From the free-electron dispersion $\varepsilon^0_{\vec{q}} = \hbar^2 q^2/(2m)$, the equality:

$$\varepsilon^0_{\vec{q}} = \varepsilon^0_{\vec{q}-\vec{K}}$$

implies

$$|\vec{q}| = \left|\vec{q} - \vec{K}\right|.$$

Geometrically, this means that the endpoint of $\vec{q}$ lies on the Bragg plane perpendicular to $\vec{K}$ and bisecting the segment that joins the origin of reciprocal space to the point $\vec{K}$ (see Fig. 4.2). The condition that this equality holds only for the chosen $\vec{K}$ ensures that $\vec{q}$ lies exclusively on that Bragg plane and not on any other plane associated with different reciprocal lattice vectors.

Therefore:

- the case of **two nearly degenerate levels** applies to an electron whose wave vector is close to satisfying the condition for a **single** Bragg scattering event;
- the case of **many nearly degenerate levels** corresponds to the treatment of a free electron state whose wave vector is near a value at which multiple **many simultaneous** Bragg reflections can occur.

Because a weak periodic potential perturbs most strongly those states lying close to Bragg planes, we now examine the solution of the two-level system. The determinant of the coefficient matrix yields the secular equation:

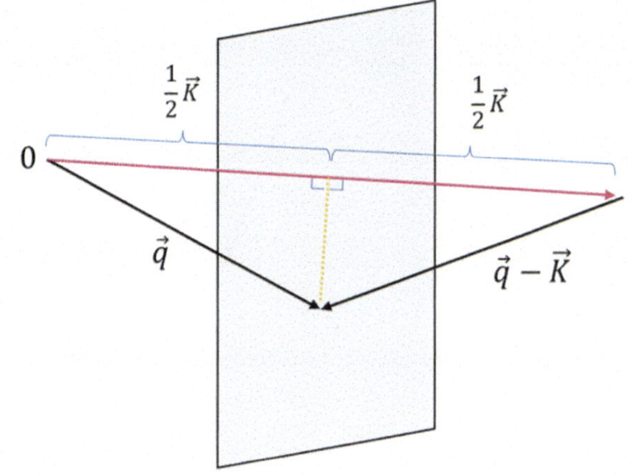

Fig. 4.2 The tips of vectors $\vec{q}$ and $\vec{q}-\vec{K}$ (for which $|\vec{q}| = \left|\vec{q}-\vec{K}\right|$) lie on the Bragg plane that bisects the line connecting the origin of $\vec{k}$-space with the reciprocal lattice point identified by the tip of the vector $\vec{K}$

$$\begin{vmatrix} \varepsilon - \varepsilon^0_{\vec{q}} & -U_{\vec{K}} \\ -U^*_{\vec{K}} & \varepsilon - \varepsilon^0_{\vec{q}-\vec{K}} \end{vmatrix} = 0,$$

which leads to

$$\left(\varepsilon - \varepsilon^0_{\vec{q}}\right)\left(\varepsilon - \varepsilon^0_{\vec{q}-\vec{K}}\right) = \left|U_{\vec{K}}\right|^2$$

and

$$\varepsilon_{1,2} = \frac{1}{2}\left(\varepsilon^0_{\vec{q}} + \varepsilon^0_{\vec{q}-\vec{K}}\right) \pm \left[\left(\frac{\varepsilon^0_{\vec{q}} - \varepsilon^0_{\vec{q}-\vec{K}}}{2}\right)^2 + \left|U_{\vec{K}}\right|^2\right]^{\frac{1}{2}}.$$

These two eigenvalues represent the leading-order modification of the free-electron energies $\varepsilon^0_{\vec{q}}$ and $\varepsilon^0_{\vec{q}-\vec{K}}$, particularly when $\vec{q}$ is close to the Bragg plane determined by $\vec{K}$ (Fig. 4.3).

By examining the energy band structure for wave vectors $\vec{q}$ parallel to $\vec{K}$, we observe that as $\vec{q}$ approaches the Bragg plane (defined by the condition $\vec{q} = {}^1\!/_2\vec{K}$) the two bands split and a band gap of magnitude $2|U_{\vec{K}}|$ opens. Conversely, when $\vec{q}$ is far from the Bragg plane, the two energy levels remain very close to their free-electron values. Exactly on the Bragg plane, the free-electron energies satisfy the degeneracy condition:

$$\varepsilon^0_{\vec{q}} = \varepsilon^0_{\vec{q}-\vec{K}}.$$

In this case, the solutions of the two-level system reduce to:

$$\varepsilon_{1,2} = \varepsilon^0_{\vec{q}} \pm \left|U_{\vec{K}}\right|$$

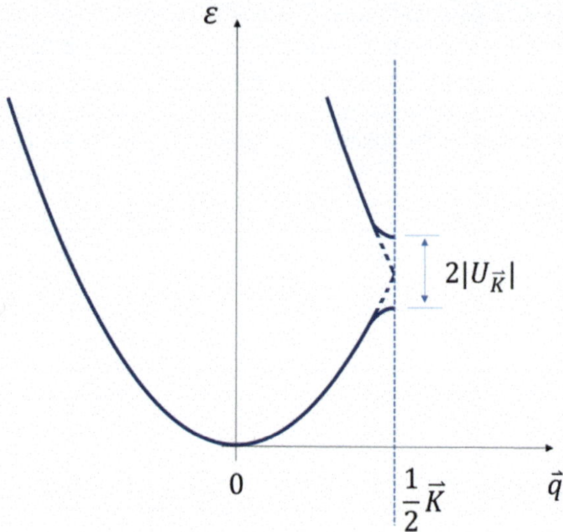

Fig. 4.3 Energy gap near the Bragg plane: the two bands become separated by a band gap of magnitude $2|U_{\vec{K}}|$

and the derivative of the energy with respect to $\vec{q}$ is given by:

$$\frac{\partial \varepsilon}{\partial \vec{q}} = \frac{\hbar 2}{m}\left(\vec{q} - \frac{1}{2}\vec{K}\right).$$

This implies that, when $\vec{q}$ lies on the Bragg plane, the gradient of $\varepsilon(\vec{q})$ is parallel to the plane itself. Points of this type, where the group velocity vanishes along the direction normal to the Bragg plane, are known as **inversion points**.

4.5 Band Structure in a 1-Dimensional Lattice

All the physical information of a one-dimensional lattice system is contained within the **first Brillouin zone**. Because the system is confined within a finite "box", its energy spectrum is discrete, and each level can be labelled by an integer n. Assuming the parabolic free-electron dispersion relation:

$$\varepsilon_n = \frac{\hbar^2 k^2}{2m},$$

we can employ the nearly-free electron approximation. In this picture, the energy bands of the solid emerge naturally from the interplay between the free-electron dispersion and the periodicity of the lattice; they are therefore a direct consequence of the confinement of the system and the imposed lattice symmetry. The band structure of a one-dimensional lattice can be represented using several equivalent schemes, typically grouped into three main categories:

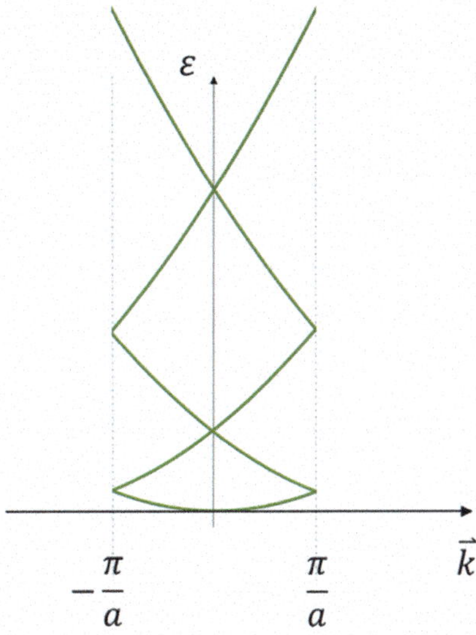

Fig. 4.4 Reduced zone scheme in a one-dimensional lattice. All energy bands are folded into the first Brillouin zone (between $-\pi/a$ and π/a)

- **Reduced zone scheme**: all energy bands are folded into the first Brillouin zone (Fig. 4.4);
- **Repeated zone scheme**: the bands are periodically replicated in each Brillouin zone (Fig. 4.5);
- **Extended zone scheme**: each band occupies a distinct Brillouin zone (Fig. 4.6).

The reduced zone scheme provides the most compact and conceptually transparent representation, as it encapsulates all the physical information without redundancy. To construct this scheme (Fig. 4.4), one proceeds as follows:

1. translate the portion of the free-electron parabola lying in the second Brillouin zone back into the first zone by subtracting a reciprocal lattice vector of magnitude $|\vec{G}|$;
2. translate the portion lying in the third Brillouin zone into the first one by subtracting $2\vec{G}$, and so on;
3. label each branch obtained from these translations with a discrete band index n.

Depending on the purpose of the analysis, any of the three schemes may be the most convenient representation. Let us now consider a complete set of wavefunctions describing the Hamiltonian H:

$$\psi_{n,\vec{k}}(\vec{r}) = e^{i\vec{k}\cdot\vec{r}} u_{n,\vec{k}}(\vec{r}),$$

where $u_{n,\vec{k}}(\vec{r})$ is periodic in direct space. Now, take another vector:

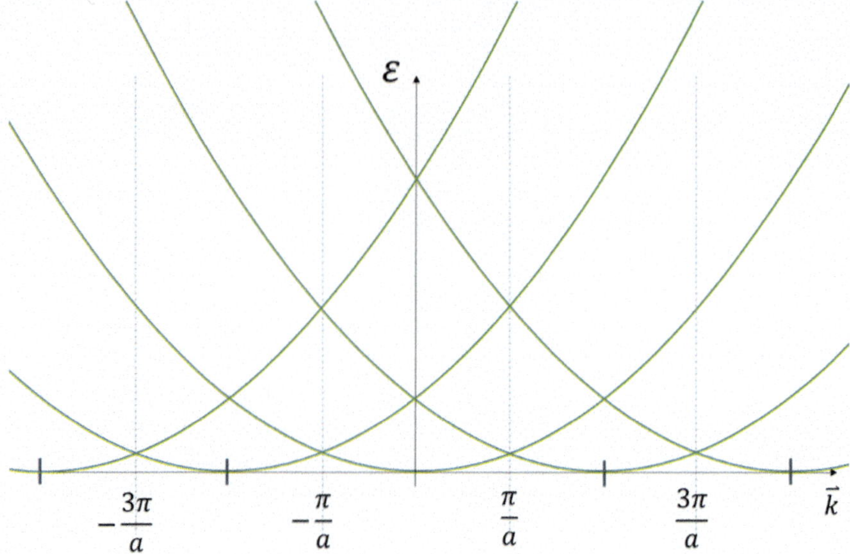

Fig. 4.5 Repeated zone scheme. The bands are periodically replicated in each Brillouin zone

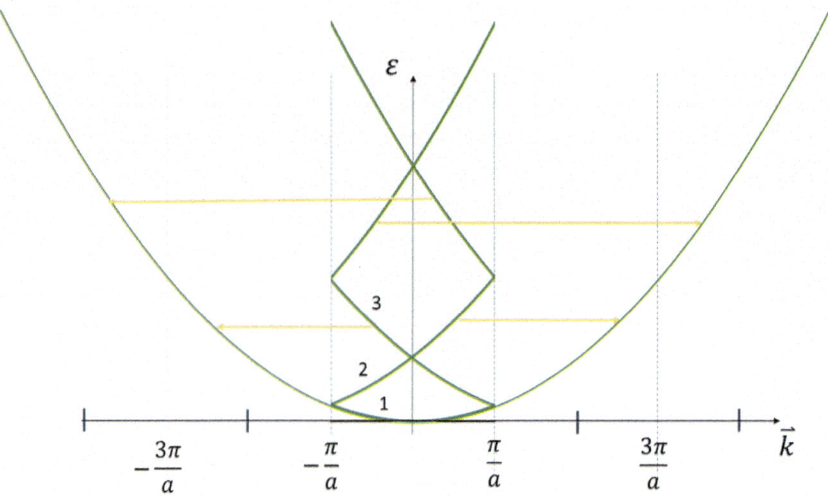

Fig. 4.6 Extended zone scheme. Each band occupies a distinct Brillouin zone

$$\vec{k}' = \vec{k} + \vec{G},$$

where $\vec{G}$ is a specific reciprocal lattice vector. We know that it is always possible to reduce the description back to the first Brillouin zone. Thus, the wavefunction $\psi_{n,\vec{k}'}(\vec{r})$ can be expressed as:

$$\psi_{n,\vec{k}'}(\vec{r}) = e^{i\vec{k}'\cdot\vec{r}} u_{n,\vec{k}'}(\vec{r}) = e^{i\vec{k}\cdot\vec{r}} e^{i\vec{G}\cdot\vec{r}} u_{n,\vec{k}'}(\vec{r}).$$

Defining the term:

$$U(\vec{r}) = e^{i\vec{G}\cdot\vec{r}} u_{n,\vec{k}'}(\vec{r}),$$

we now show that it is necessarily periodic in $\vec{r}$, so in the direct space. To verify that $U(\vec{r}) = U\left(\vec{r} + \vec{R}\right)$, consider a translation in direct space by a lattice vector $\vec{R}$, so that $\vec{r}' = \vec{r} + \vec{R}$:

$$\psi_{n,\vec{k}'}(\vec{r}') = \psi_{n,\vec{k}'}\left(\vec{r} + \vec{R}\right) = e^{i\vec{k}\cdot\vec{r}} e^{i\vec{G}\cdot\vec{r}} e^{i\vec{G}\cdot\vec{R}} u_{n,\vec{k}'}\left(\vec{r} + \vec{R}\right).$$

Using the property $e^{i\vec{G}\cdot\vec{R}} = 1$ and the periodicity of $u_{n\vec{k}'}(\vec{r}) = u_{n,\vec{k}'}\left(\vec{r} + \vec{R}\right)$:

$$\psi_{n,\vec{k}'}\left(\vec{r}'\right) = e^{i\vec{k}\cdot\vec{r}} e^{i\vec{G}\cdot\vec{r}} u_{n,\vec{k}'}(\vec{r}),$$

we obtain:

$$\psi_{n,\vec{k}'}(\vec{r}') = e^{i\vec{k}\cdot\vec{r}} U(\vec{r}).$$

Therefore, Bloch states labelled by wave vectors differing by a reciprocal lattice vector represent the same physical eigenstate, although their wavefunctions differ by a position-dependent phase factor: shifting $\vec{k}$ by a reciprocal lattice vector produces an equivalent state within the first Brillouin zone.

4.6 Born-Von Karman Boundary Conditions

To derive a condition that constrains the possible values of the wave vector $\vec{k}$ and ensures that it remains real, it is necessary to impose a boundary condition on the wave function. This condition enforces periodicity of the wave function within a given Bravais lattice. Consider first a one-dimensional lattice of total length L, consisting of N identical cells, each of length a, as illustrated in Fig. 4.7.

To guarantee that the system exhibits bulk properties, the wavefunction must take the same value at both ends of the crystal. A convenient way to implement this requirement is to imagine "closing" the system into a circular geometry whose circumference is equal to L. This circular configuration (Fig. 4.8) contains the same number N of cells, each of length a, so that $L = Na$.

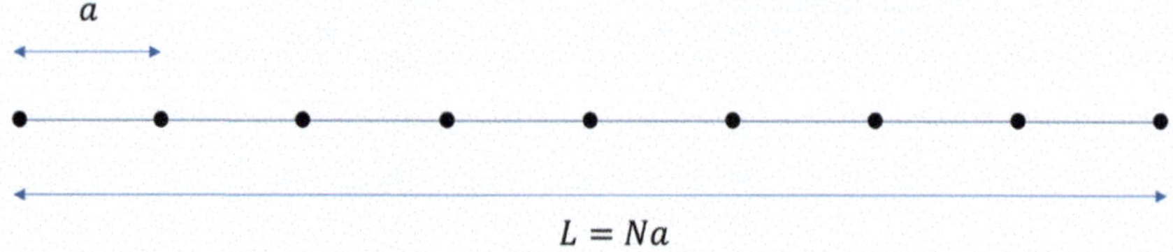

Fig. 4.7 One-dimensional lattice of total length L, consisting of N identical cells, each of length a

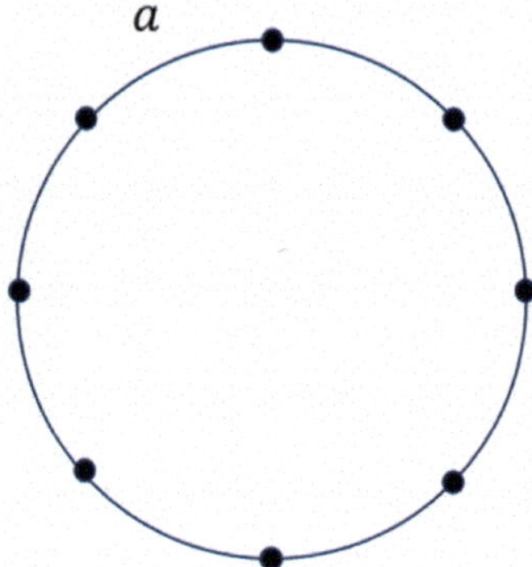

Fig. 4.8 One-dimensional lattice of length L in a circular shape, with N cells having length a

The periodic boundary condition can then be written as:

$$\psi(r + Na) = \psi(r).$$

This idea generalizes naturally to three dimensions:

$$\psi(\vec{r} + N_i\vec{a}_i) = \psi(\vec{r}),$$

where the integers N_i ($i = 1, 2, 3$) are extremely large, of order $N^{1/3}$, with N the number of lattice sites (where $N = N_1N_2N_3$). As shown in Fig. 4.9 an analogous construction can be visualized for a three-dimensional cubic lattice, in which opposite faces are joint by "closing" the cube within a higher-dimensional embedding space (four-dimensional space).

Since the wavefunction must be periodic at the boundaries of the crystal, the Bloch function must satisfy:

$$\psi_{n,\vec{k}}\left(\vec{r} + \vec{R}\right) \equiv \psi_{n,\vec{k}}(\vec{r}).$$

By applying both the translation operator $e^{i\vec{k}\cdot\vec{R}}$ and the Bloch theorem to the wavefunction, where $\vec{R} = \sum_{i=1}^{3} N_i\vec{a}_i$, we obtain:

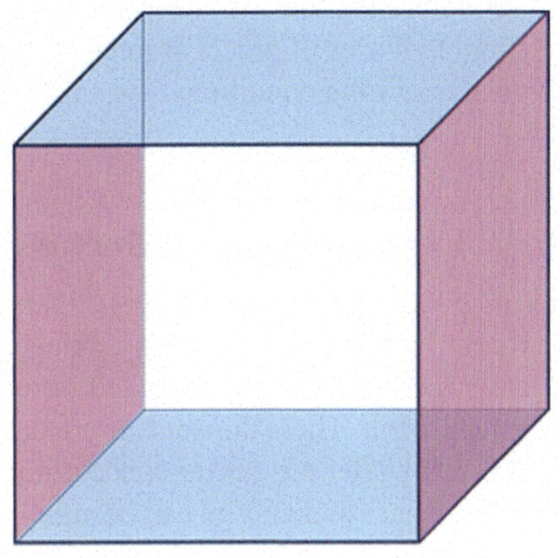

Fig. 4.9 Three-dimensional lattice. The opposite sides of the cube (i.e., the pink faces, the light blue faces, and the white ones) can be overlapped by closing the cube on itself in a four-dimensional space

$$\psi_{n,\vec{k}}\left(\vec{r}+\vec{R}\right) = e^{i\vec{k}\cdot\vec{R}}\psi_{n,\vec{k}}(\vec{r}) = e^{i\vec{k}\cdot\left(\sum_{i=1}^{3} N_i\vec{a}_i\right)}\psi_{n,\vec{k}}(\vec{r}) \equiv \psi_{n,\vec{k}}(\vec{r}).$$

Thus:

$$e^{i\vec{k}\cdot\sum_{i=1}^{3} N_i\vec{a}_i} = 1.$$

This leads directly to the quantization condition:

$$\vec{k}\cdot\sum_{i=1}^{3} N_i\vec{a}_i = 2\pi\sum_{i=1}^{3} n_i,$$

where $n_i \in Z$. Given that $\vec{k}$ is a reciprocal lattice vector, it can be expanded in the reciprocal lattice basis as:

$$\vec{k} = \sum_{j=1}^{3} x_j\vec{b}_j,$$

without any additional constraints. Substituting this expression in the previous equation and using the relation:

$$\vec{a}_i\cdot\vec{b}_j = 2\pi\delta_{ij}$$

we derive:

$$\sum_{i,j=1}^{3} N_i x_j \vec{a}_i\cdot\vec{b}_j = 2\pi\sum_{i=1}^{3} N_i x_i = 2\pi\sum_{i=1}^{3} n_i.$$

Therefore, the coefficients x_i must be a rational of the form:

$$x_i = \frac{n_i}{N_i}$$

and the allowed values of $\vec{k}$ that can be used in the Bloch function are:

$$\vec{k} = \sum_{i=1}^{3} \frac{n_i}{N_i} \vec{b}_i,$$

where the integers n_i label the discrete allowed states.

4.7 What is the Minimum Allowed Volume for $\vec{k}$-state?

To determine the minimum volume associated with each $\vec{k}$-state, we start by considering the quantity:

$$\Delta\vec{k} = \frac{\vec{b}_1}{N_1} \cdot \frac{\vec{b}_2}{N_2} \times \frac{\vec{b}_3}{N_3} = \frac{\vec{b}_1 \cdot \vec{b}_2 \times \vec{b}_3}{N},$$

where $\vec{b}_i/N_i$ $(i = 1, 2, 3)$ represents the reciprocal lattice vector components scaled by the number of lattice points N_i in the respective crystallographic directions. Here, $v = \vec{b}_1 \cdot \vec{b}_2 \times \vec{b}_3$ denotes the volume of the first Brillouin zone, and $N = N_1 N_1 N_3$ is the total number of lattice cells in the direct lattice. For each electronic band, a maximum of $2N$ electrons can be accommodated due to spin degeneracy.

Now, given a crystal with a number t of electrons per lattice point, we can determine how many completely filled bands exist by dividing the total number of electrons (tN) by the number of electrons that each band can hold ($2N$):

$$\frac{tN}{2N} = \frac{t}{2}.$$

From this expression, two possible scenarios arise determined by the parity of t:

1. if t is **odd**, then $t/2 \notin Z$. In this scenario, certain energy bands are **partially filled**. This incomplete filling necessitates metallic behavior, as the system exhibits a continuum of available energy states near the Fermi level;
2. if t is **even**, then $t/2 \in Z$. The system can be further categorized into two distinct electronic subcases:

   i. **semiconductors/insulators**: the bands are separated by an energy gap (or band gap). Consequently, the bands are either completely filled or completely empty, resulting in no partially filled bands;

ii. **metals**: the energy bands may overlap. This overlap permits the partial filling of bands even when $t/2$ is an integer. Due to this overlap, electrons occupy lower energy states first, resulting in partially filled bands.

In conclusion, the electronic structure of the crystal, and thus its classification as a metal, semiconductor, or insulator, is fundamentally dependent on the parity of t and the presence or absence of band overlap.

References

1. Ashcroft NW, Mermin ND (1976) Solid state physics. Holt, Rinehart and Winston, New York
2. Kittel C (2005) Introduction to solid state physics, 8th edn. Wiley, Hoboken
3. Ziman JM (1972) Principles of the theory of solids, 2nd edn. Cambridge University Press, Cambridge
4. Madou MJ, Morrison SR (1997) Chemical sensing mechanisms in solid-state devices. In: Chemical sensing with solid state devices. Academic Press, pp 74–83
5. Harrison WA (1980) Solid state theory. Dover Publications, New York
6. Blatt FJ, Weisskopf VF (1979) Theoretical nuclear physics: an introduction. Springer, New York
7. Sakurai JJ, Napolitano J (2017) Modern quantum mechanics, 2nd edn. Cambridge University Press, Cambridge
8. Cohen-Tannoudji C, Diu B, Laloë F (1977) Quantum mechanics, vols I–II. Wiley, New York
9. Wigner E, Seitz F (1933) On the constitution of metallic sodium. Phys Rev 43:804
10. Bloch F (1929) Über die Quantenmechanik der Elektronen in Kristallgittern. Z Phys 52:555–600
11. Yu PY, Cardona M (1996) Electronic band structures. In: Fundamentals of semiconductors. Springer, Heidelberg
12. Slater JC (1937) Wave functions in a periodic potential. Phys Rev 51:846
13. Press WH (1992) Numerical recipes. Cambridge University Press, New York
14. Keller HB (1968) Numerical methods for two-point boundary-value problems. Blaisdell, Massachusetts

Chapter 5
Semiconductors

> Before I came here I was confused about this subject. Having listened to your lecture I am still confused. But on a higher level. (Enrico Fermi)

This Chapter is devoted to semiconductors, which are key materials in modern electronics and optoelectronics. Their understanding requires a detailed analysis of the quantum properties of electrons and of the band structure. The Chapter begins by introducing the concept of Fermi energy in one-, two-, and three-dimensional systems, as well as the role of the Fermi level as the electrochemical potential of electrons. It then discusses the main types of semiconductors, emphasizing the importance of the effective mass in charge carrier dynamics. The treatment continues with the analysis of non-degenerate semiconductors, focusing on the distinction between intrinsic and extrinsic materials, and concludes with the study of band bending, a phenomenon that plays a central role in semiconductor junctions and device operation. Building upon the literature cited here [1–18], the authors provide an original and integrated interpretation, particularly regarding the Fermi level as the electrochemical potential of electrons, as illustrated in Malagù's poster at the International Meeting on Chemical Sensors (IMCS), held in Freiburg from June 22 to 26, 2025 [19].

5.1 Fermi Energy and Fermi Level

Let us consider an energy eigenvalue within a band structure as a function of the wave vector $\vec{k}$. At absolute zero temperature ($T = 0$ K), there exists a maximum energy value, denoted by ε_F, such that, according to the Pauli exclusion principle, all quantum states with energies below ε_F are fully occupied, while those with energies above ε_F are completely empty. This quantity is known as **Fermi energy**, defined as the energy of the highest occupied quantum state in a system of fermions at absolute zero temperature. It therefore acts as a sharp boundary separating occupied from unoccupied states.

© The Author(s), under exclusive license to Springer Nature Switzerland AG 2026

C. Malagù and G. Zonta, *Gas Sensors*, https://doi.org/10.1007/978-3-032-21614-4_5

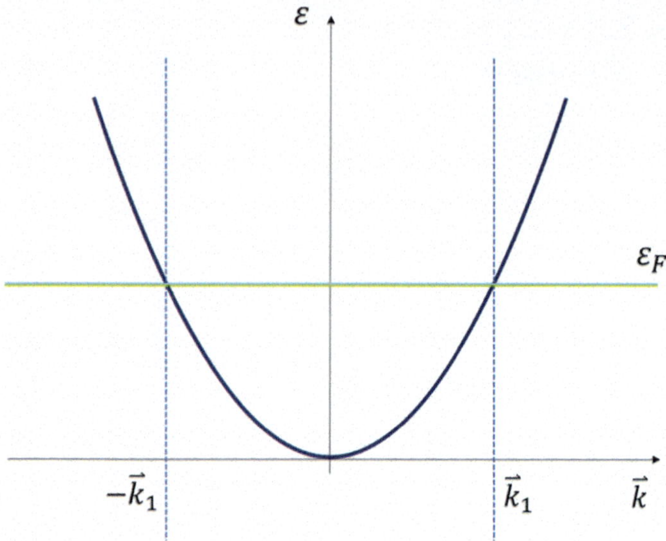

Fig. 5.1 Fermi energy in one dimension

In one dimension (1D), the Fermi energy is given by (see Fig. 5.1):

$$\varepsilon_F = \frac{\hbar^2 k^2}{2m}.$$

In two dimensions (2D), the Fermi energy depends on the components k_x and k_y of the wave vector $\vec{k}$:

$$\varepsilon_F = \frac{\hbar^2}{2m}\left(k_x^2 + k_y^2\right).$$

This expression describes a circle in the k_x–k_y plane. Writing it as:

$$k_x^2 + k_y^2 = \frac{2m\varepsilon_F}{\hbar^2},$$

we can interpret the right-hand side as a squared radius R:

$$k_x^2 + k_y^2 = R^2,$$

expressed as:

$$R = \sqrt{\frac{2m\varepsilon_F}{\hbar^2}}.$$

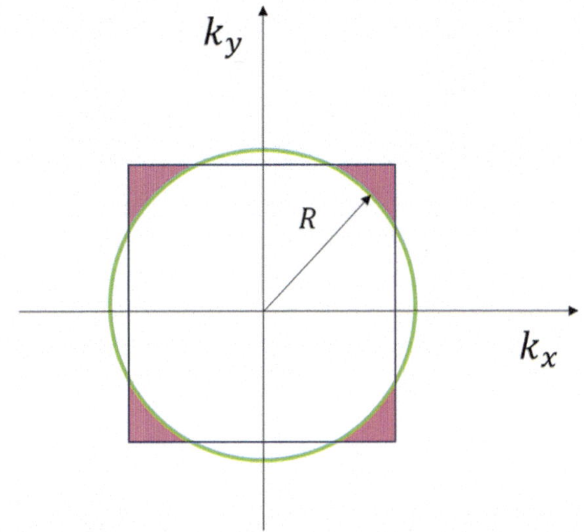

Fig. 5.2 Fermi energy in two dimensions, depicted by the Fermi circle (in green) with radius R. The first Brillouin zone is outlined by the square, with the regions not occupied by electrons highlighted in pink

As shown in Fig. 5.2, we obtain the **Fermi circle**: states inside the circle are occupied, whereas those outside are unoccupied.

Similarly, in three dimensions (3D), the Fermi energy becomes:

$$\varepsilon_F = \frac{\hbar^2}{2m}\left(k_x^2 + k_y^2 + k_z^2\right),$$

which defines the **Fermi sphere** (Fig. 5.3).

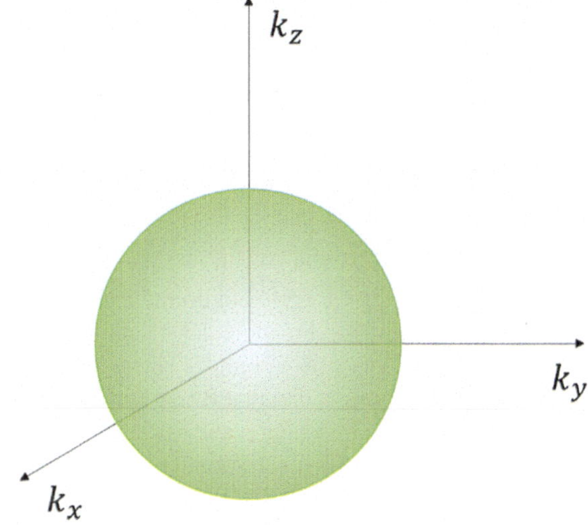

Fig. 5.3 Fermi energy in three dimensions, represented with the Fermi sphere

Being $\vec{k}_F$ the Fermi-sphere radius, the volume of the Fermi sphere in the $\vec{k}$-space is given by:

$$V_{\text{sphere}} = \frac{4}{3}\pi k_F^3.$$

The total number of electrons (N_{el}) within the volume of the Fermi sphere can be determined by dividing its volume by the minimum allowed volume for a $\vec{k}$-state (Δk) and multiplying by 2 to account for spin degeneracy (since each $\vec{k}$-state can accommodate a maximum of two electrons). This can be expressed as:

$$N_{el} = 2\frac{V_{\text{sphere}}}{\Delta k} = 2\frac{\frac{4}{3}\pi k_F^3}{\Delta k}.$$

Knowing that the first Brillouin zone volume is given by:

$$v_{IBZ} = \frac{(2\pi)^3}{v},$$

we can calculate the minimum allowed volume for a $\vec{k}$-state as:

$$\Delta k = \frac{v_{IBZ}}{N} = \frac{(2\pi)^3}{vN} = \frac{(2\pi)^3}{V}.$$

where N is the total number of allowed $\vec{k}$-states, v the volume of the primitive cell, and V the total crystal volume.

The equation can be rearranged to find the electron density n as:

$$n = \frac{N_{el}}{V} = \frac{4}{3}\pi k_F^3 \frac{2}{(2\pi)^3},$$

that becomes:

$$n = \frac{k_F^3}{3\pi^2}.$$

In general, this formula can be extended to compute the electron density for arbitrary volumes in $\vec{k}$-space ($V_{\vec{k}-\text{space}}$), other than the Fermi sphere one:

$$n = \frac{2}{(2\pi)^3} V_{\vec{k}-\text{space}},$$

where:

$$V_{\vec{k}-\text{space}} = \int_{\text{band}} f_n[E(\vec{k})]d^3k,$$

where f_n is a generic electron distribution function of energy $E(\vec{k})$ and the integral is performed over a specific band (e.g., conduction or valence band). Using this volume, it is so possible to find the electron distribution n.

For electrons, the distribution function is given by the **Fermi-Dirac** distribution:

$$f_n(E) = \frac{1}{1 + e^{\frac{E-E_F}{K_BT}}}$$

where E_F is the **Fermi Level**, which must be distinguished from the Fermi energy. Mathematically, when $E = E_F$, the probability of occupation is exactly $1/2$:

$$f_n(E) = \frac{1}{2}.$$

At $T = 0$ K, the distribution reduces to a step function:

$$f_n(E) = \begin{cases} 0, & \text{if } E > E_F \\ 1, & \text{if } E < E_F \end{cases}.$$

To illustrate the concept further, consider the infinite alternating series:

$$S_n = \sum_{n=1}^{\infty}(-1)^n = -1 + 1 - 1 + 1 - 1 + 1 - \dots$$

This series behaves such that:

$$S_n = \begin{cases} 0, & \text{if } n \text{ is even} \\ -1, & \text{if } n \text{ is odd.} \end{cases}$$

The series can be associated with its average value, which is $-1/2$. This analogy helps to conceptualize how the Fermi–Dirac distribution approaches a value of $1/2$ at $E = E_F$. This analogy is purely illustrative and does not replace the statistical-mechanical derivation. Figure 5.4 represents graphically the Fermi distribution for two scenarios: $T = 0$ K (green line) and $T \neq 0$ K (blue line).

In the first case (at $T = 0$ K), for energies $E > E_F$, the probability of finding an electron is zero, while for $E < E_F$, it is unity. This behavior is analogous to the modulus of series S_n discussed earlier. When the temperature is raised above absolute zero ($T \neq 0$ K), the distribution becomes continuous, and deviations from E_F are measured in units of K_BT. As temperature increases, so does the probability of finding electrons at energy levels higher than E_F.

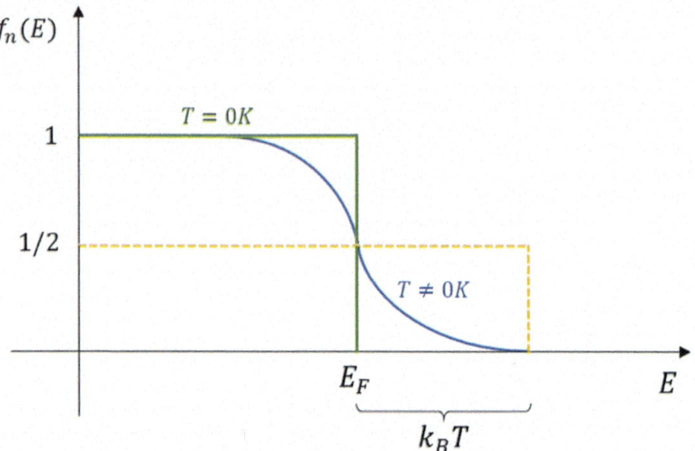

Fig. 5.4 Fermi–Dirac distribution value $f_n(E)$ as a function of energy E (where E_F is the Fermi level) at $T = 0$ K (green line) and at $T \neq 0$ K (blue line)

This visualization clarifies the meaning of the Fermi level: at finite temperature, when $E = E_F$, the occupation probability is exactly $1/2$. Thus, the Fermi level corresponds to the energy for which there is a 50% probability of occupation.

In the limit $\rightarrow 0$ K, the Fermi–Dirac distribution approaches a step function in which the occupation probability takes only the values 1 or 0. Nonetheless, even in this limit, the probability at $E = E_F$ remains exactly $1/2$. This ensures that the notion of the Fermi level remains meaningful at absolute zero. Crucially, the Fermi level is well defined for all classes of materials (metals, semiconductors, and insulators) because its interpretation is intrinsically probabilistic. In contrast, the Fermi energy loses its physical relevance in insulators and semiconductors, where it no longer identifies the highest occupied energy at absolute zero temperature.

For a Fermi sphere, the Fermi energy is defined as:

$$\varepsilon_F = \frac{\hbar^2 k_F^2}{2m},$$

indicating that all states inside the Fermi sphere are fully occupied, whereas all states outside are empty. More generally, the **Fermi surface** encompasses all points where $\varepsilon_n\left(\vec{k}\right) = \varepsilon_F$. This is analogous to an equipotential surface, where every point shares the same potential, but in this context, it represents all points with the same degenerate energy eigenvalue at $T = 0$ K.

We now turn to several examples illustrating how the Fermi surface can be used to distinguish metallic from insulating behavior. The distinction between metals and non-metals depends on the parameter $t/2$, depending on whether it is an integer (as previously discussed in Sect. 4.7). However, even when $t/2$ is an integer, a crystal may behave as either a metal or a non-metal depending on whether band overlap occurs.

The first example, a straightforward case, concerns the **Group I** elements (**alkali metals**), such as Li, Na, and K, which crystallize in a body-centered cubic structure (see Fig. 3.3). These elements have a single valence electron ($t = 1$), so the unit-cell volume is twice that of the primitive cell (see Sect. 3.1). The electron density is therefore $n = 2/a^3$ (two electrons per unit cell). Equating this to $n = k_F^3/(3\pi^2)$, where k_F is the radius of the Fermi sphere and $2\pi/a$ is the size of the first Brillouin zone, yields:

$$k_F^3 = 2\frac{3\pi^2}{a^3} = 2\frac{3\pi^2}{a^3}\frac{(4\pi)}{(4\pi)} = \frac{8\pi^3}{a^3}\frac{3}{4\pi} = \left(\frac{2\pi}{a}\right)^3\frac{3}{4\pi}.$$

It follows that:

$$k_F \sim 0.6\frac{2\pi}{a}.$$

Numerical simulations show that the minimum distance from the origin of $\vec{k}$-space to a face of the first Brillouin zone is approximately:

$$d \sim 0.72\frac{2\pi}{a}.$$

Hence, the Fermi sphere is entirely contained within the first Brillouin zone (see Fig. 5.5). The Fermi surface is therefore nearly spherical, and the valence electrons can be treated as free electrons, so the crystal exhibits metallic behavior.

If the Fermi sphere were larger, it would touch or intersect a Bragg plane. Since the energy eigenvalues near a Bragg plane are modified by the lattice potential at first order, the Fermi sphere becomes distorted whenever it touches or crosses such a plane.

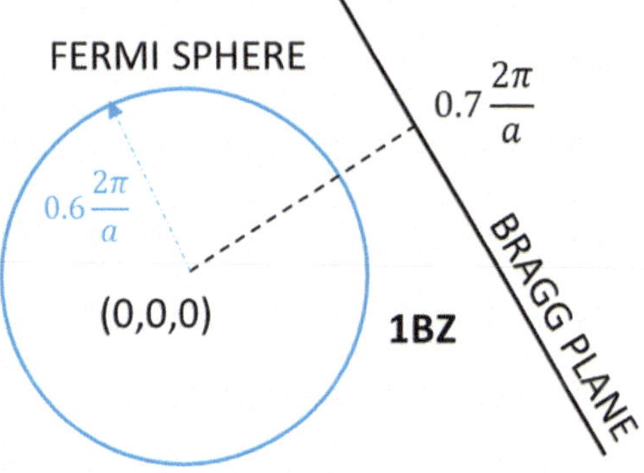

Fig. 5.5 Fermi sphere of alkali metals, contained into the first Brillouin zone (1BZ)

Now, let us examine the second, non-trivial case of **alkaline-earth metals (Group II)**, such as Be, Mg, Ca, Sr, and Ba, each of which contributes two valence electrons per lattice site. These elements also crystallize in cubic structures, but their valence is $t = 2$ (they are in the second column of the periodic table). The volume of the first Brillouin zone (1BZ), where v is the primitive-cell volume and V is the total crystal volume, is:

$$\vec{b}_1 \cdot \vec{b}_2 \times \vec{b}_3 = \frac{(2\pi)^3}{v} = \frac{(2\pi)^3 N}{V} = \frac{2N\left(4\pi^3\right)}{V} = n4\pi^3$$

where n is the electron density. For metals, we also know that $n = k_F^3/3\pi^2$, so the volume of the Fermi sphere is:

$$\frac{4\pi}{3}k_f^3 = \frac{4\pi}{3}\left(n3\pi^2\right) = n4\pi^3.$$

Thus, the volume of the 1BZ matches that of the Fermi sphere. In this situation, Bragg planes may intersect the Fermi sphere in some regions. If the lattice potential is sufficiently strong, it can distort the Fermi sphere until it coincides with the shape of the 1BZ. In this scenario, the Fermi surface disappears, the band becomes completely filled, and the material behaves as an insulator. Conversely, if the potential is not strong enough to deform the Fermi surface sufficiently to eliminate it, parts of the 1BZ will intersect the Fermi sphere, and the crystal will remain metallic even though the valence is even.

We now examine the significance of the Fermi level in insulators and semiconductors. Its physical meaning is tied to its position relative to the conduction-band minimum (CBM) and the valence-band maximum (VBM). Metals will be excluded from consideration for the moment. Semiconductors may generally be divided into two categories:

- **intrinsic semiconductors**: these possess a small band gap, are undoped, and at $T = 0$ K have their Fermi level located at the midpoint of the gap ($E_F = 1/2(E_C + E_V)$);
- **extrinsic semiconductors**: these are doped with donor impurities (n-type), which shift E_F toward the conduction band, or acceptor impurities (p-type), which shift E_F toward the valence band.

In extrinsic semiconductors, doping can shift the Fermi level into the conduction band (n-type) or the valence band (p-type). In such cases, the material exhibits metallic behavior, and the Fermi level assumes the same physical role as the Fermi energy, separating filled from empty states. This occurs when $E_F > E_{\text{CBM}} + 3k_BT$ for n-type materials or $E_F < E_{\text{VBM}} - 3k_BT$ for p-type materials. The semiconductor cannot be degenerate at $T = 0$ K. The metallic behavior of a degenerate semiconductor implies that $n = k_F^3/\left(3\pi^2\right)$, as in ordinary metals. Thus, while the Fermi level and the Fermi energy are related, their meanings are not generally identical: the Fermi level retains a broad probabilistic interpretation, and only in metals and degenerate semiconductors do the two concepts coincide.

5.2 Fermi Level: Electrochemical Potential of Electrons

Let us now examine the Fermi–Dirac distribution, focusing in particular on the Fermi level E_F appearing in the exponential term of the denominator. In this context, the Fermi level acquires a purely probabilistic interpretation and can be identified with the **electrochemical potential of electrons** ($E_F = \overline{\mu}^e$).

We begin by recalling some principles of classical thermodynamics:

- **Zeroth Law of Thermodynamics:** "*If two bodies A and B are each in thermal equilibrium with a third body C, then A and B are also in thermal equilibrium with one another*".

  This principle implies that two systems in thermal equilibrium share the same temperature. From an intuitive perspective, the Zeroth Law respects symmetry (if $T_A = T_B$, so $T_B = T_A$) and reflexivity ($T_A = T_A$). Additionally, it introduces the concept of transitivity, allowing thermal equilibrium to be treated as an equivalence relation. As a consequence, states of thermal equilibrium can be grouped into equivalence classes. Absolute temperature T is the physical quantity characterizing these classes. For instance, if $T_A = T_C$ and $T_B = T_C$, then $T_A = T_B$, meaning A and B are in thermal equilibrium.

- **First Law of Thermodynamics:** "*The internal energy of an isolated thermodynamic system remains constant*".

  This principle is mathematically formulated as:

$$\delta Q - dU + \delta L = dU + pdV,$$

or, equivalently:

$$dU = \delta Q - \delta L.$$

Here: δQ is the heat exchanged by the system, positive ($\delta Q > 0$) for adsorption and negative ($\delta Q < 0$) for release; δL is the infinitesimal work performed, positive ($\delta L > 0$) when done by the system and negative ($\delta L < 0$) when done on the system; dU is the infinitesimal variation of the internal energy, a state function. In the specific case $\delta Q = \delta L$ (both in joules in SI units), the system is isolated both thermally and mechanically, resulting in no change in internal energy ($dU = 0$). One significant consequence of this principle is the *conservation of the total energy of the system.*

- **Second Law of Thermodynamics:** "*In an insulated system the entropy cannot decrease*".

  Heat exchange is described by the inexact differential δQ, whose value depends on the path of integration. This reflects the fact that the exchanged heat varies depending on whether the process occurs under constant temperature or constant pressure. Entropy is defined by:

$$\left.\frac{\delta Q}{T}\right|_{\text{Rev}} = dS$$

where dS is an exact differential for reversible processes. As a state function, entropy is integrable and depends only on the initial and final states. The Second Law of Thermodynamics can therefore be written as:

$$dS \geq \frac{\delta Q}{T},$$

with equality holding only for reversible processes. For real (irreversible) processes, which include dissipation such as friction, the inequality becomes strict:

$$dS > \frac{\delta Q}{T}.$$

Thus, during an irreversible transformation:

$$\Delta S > \int \frac{\delta Q}{T}.$$

Additionally, the Second Law provides a definition of absolute temperature:

$$T = \left.\frac{\delta Q}{dS}\right|_{\text{Rev}}.$$

This formulation leads to intriguing phenomena in certain physical systems that challenge the conventional assumption $T \geq 0$. For instance, in quantum systems with spin degrees of freedom, negative temperatures $T < 0$ may arise. In such systems, if heat is absorbed ($\delta Q > 0$), the entropy decreases ($dS < 0$), meaning the system becomes "more ordered". Remarkably, a system at negative temperature transfers heat to any system at positive temperature, even one at arbitrarily high T (e.g., a temperature "higher" than positive infinity). Thus, negative temperatures correspond to states "hotter" than $T = +\infty$.

- **Third Law of Thermodynamics (or Nernst Theorem):** "*For any condensed system undergoing a reversible isothermal transformation, the corresponding entropy change approaches zero as the temperature approaches* $T = 0$ K"

In other words, $S \to 0$ as $T \to 0$. Returning to the Second Law of Thermodynamics:

$$dS \geq \frac{\delta Q}{T},$$

which may be rewritten as:

$$\delta Q - TdS \leq 0,$$

let us now introduce the concept of **Enthalpy**, defined as:

$$H = U + PV,$$

where U is the internal energy, P is the pressure, and V is the volume. Since these quantities are thermodynamic potentials, they can be written as exact differentials. The differential of enthalpy is therefore:

$$dH = dU + PdV + VdP.$$

When the system undergoes a heat exchange at constant pressure ($dP = 0$), this simplifies to:

$$dH = dU + PdV,$$

which, according to the First Law of Thermodynamics, is equivalent to δQ. Hence, at constant pressure:

$$dH = \delta Q,$$

indicating that enthalpy represents the heat exchanged by a physical system under isobaric conditions.

We now define a new thermodynamic potential, the **Gibbs free energy** G, expressed as:

$$G = H - TS,$$

or, equivalently:

$$G = U + PV - TS.$$

Its differential form is:

$$dG = dU + PdV + VdP - TdS - SdT.$$

In the case of an **isobaric** ($dP = 0$) and **isothermal** ($dT = 0$) conditions, this reduces to:

$$dG = dU + PdV - TdS.$$

Using the First Law:

$$dG = \delta Q - TdS$$

and since $\delta Q - TdS \leq 0$, the **Second Law** can be rewritten in terms of the Gibbs free energy as:

$$dG \leq 0.$$

This implies that a transformation can proceed spontaneously as long as G decreases, ceasing only once a minimum of Gibbs free energy is reached.

All these considerations apply to **closed thermodynamic systems**, where the total amount of matter remains constant. Such systems may exchange heat and work with the surroundings but do not exchange particles. Therefore, the Second Law remains applicable in its original form.

For **open thermodynamic systems**, the First Law of Thermodynamics is modified as:

$$dU = TdS - PdV + \sum_{i=1}^{N} \mu_i dN_i$$

where μ_i are proportionality coefficients (later identified as chemical potentials) that ensure the consistency of dimensions, and N_i is the amount of each constituent species (e.g., interacting elements, electrons, or particles). The last term in this equation indicates the possibility of matter exchange with the external environment, which distinguishes open systems from closed or isolated systems.

Using the differential of the Gibbs free energy for processes under isothermal ($dT = 0$) and isobaric ($dP = 0$) conditions:

$$dU = dG - PdV + TdS,$$

and substituting into the expression for dU, we find:

$$dG = \sum_{i=1}^{N} \mu_i dN_i.$$

Thus, Gibbs free energy is a **first-order linear homogeneous function** of the mixture constituents, which is valid for any mixture. The Gibbs free energy of a mixture can therefore be expressed as:

$$G = \sum_{i=1}^{N} \mu_i N_i.$$

Now, considering a general homogeneous function f of degree n, it holds that:

$$f(\alpha x_1, \alpha x_2, \ldots, \alpha x_n) = \alpha^n f(x_1, x_2, \ldots, x_n).$$

From this property, the **Euler theorem** applies:

$$\sum_i \frac{\partial f}{\partial x_i} x_i = nf.$$

Since Gibbs free energy is a homogeneous function of degree one:

$$\sum_{i=1}^{N} \mu_i N_i = \sum_{i=1}^{N} \left.\frac{\partial G}{\partial N_i}\right|_{N_j} N_i,$$

where $\boldsymbol{\mu_i}$ is identified as the **chemical potential**, thermodynamically defined as the partial derivative of the Gibbs free energy with respect to the i - th constituent, while keeping all other N_j, T, and P constant:

$$\mu_i = \left.\frac{\partial G}{\partial N_i}\right|_{N_j,T,P}.$$

If charged species are involved, the condition $\Delta G \leq 0$ no longer suffices to define the point at which a transformation ceases. An additional term corresponding to **useful work** $\boldsymbol{W_U}$ must be included:

$$\Delta G + W_U \leq 0.$$

The term W_U is defined as:

$$W_U = z_i F \Delta\phi,$$

where z_i is a charge number coefficient ($z_i = +1$ for positive charges, $z_i = -1$ for negative charges, and $z_i = 0$ for neutral particles), F the Faraday constant (the charge of one mole of electrons), and $\Delta\phi$ the electrostatic potential variation.

Thus, the equation becomes:

$$\Delta G + z_i F \Delta\phi = \Delta(G + z_i F \phi) \leq 0.$$

Now, consider a simple system composed of one mole ($n = 1$) of an ideal gas, for which:

$$PV = RT.$$

From the thermodynamic definition of the Gibbs free energy $G = U + PV - TS$, it follows that:

$$\frac{\partial G}{\partial P} = V.$$

Substituting into the ideal gas law:

$$\frac{\partial G}{\partial P} = \frac{RT}{P}.$$

Separating the variables and integrating from a reference state, as:

$$\int_{G_0}^{G} \partial G = \int_{P_0}^{P} \frac{RT}{P} \partial P,$$

we obtain:

$$G = G_0 + RT \ln\left(\frac{P}{P_0}\right),$$

where G_0 is Gibbs free energy under standard conditions and P_0 is the standard pressure (commonly the atmospheric pressure).

For a single-component system, Gibbs free energy is the **chemical potential** μ:

$$\mu = G,$$

therefore:

$$\mu = \mu_0 + RT \ln\left(\frac{P}{P_0}\right).$$

If instead the system consists of more than one constituent (which can also be charged), the partial free energy differs from the chemical potential. Recalling that to stop a reaction:

$$\Delta(G + z_i F\phi) = 0,$$

this implies:

$$\Delta(\overline{G}) = 0,$$

where:

$$\overline{G} = G + z_i F\phi.$$

Thus, adding the electrostatic term $z_i F\phi$ to the Gibbs free energy leads to the **electrochemical potential** $\overline{\mu}$ for one mole:

$$\overline{\mu} = \mu_0 + RT \ln\left(\frac{P}{P_0}\right) + z_i F\phi,$$

which is the sum of the chemical potential and the contribution of the electrostatic potential. Dividing this expression by Avogadro's number N_A:

$$\frac{\overline{\mu}}{N_A} = \frac{\mu_0}{N_A} + \frac{R}{N_A} T \ln\left(\frac{P}{P_0}\right) + z_i \frac{F}{N_A} \phi$$

and using the Boltzmann constant $k_B = R/N_A$ and the electron charge $e = F/N_A$, the equation becomes:

$$\frac{\overline{\mu}}{N_A} = \frac{\mu_0}{N_A} + k_B T \ln\left(\frac{P}{P_0}\right) + z_i e \phi.$$

The**electrochemical potential for an electron**can then be expressed as:

$$\overline{\mu}^e = \mu_0^e + k_B T \ln\left(\frac{P}{P_0}\right) - e\phi,$$

or equivalently (being $z_i = -1$ for an electron):

$$\overline{\mu}^e = \mu - e\phi,$$

where μ is the chemical potential.

This directly links to the **Fermi-Dirac distribution**, which is given by:

$$f_n(E) = \frac{1}{1 + e^{\frac{E - \overline{\mu}}{k_B T}}}.$$

In electrochemistry, the electrochemical potential corresponds exactly to the **Fermi level** in solid-state physics. Thus, from here, we will refer to the Fermi level as the electrochemical potential of electrons.

5.3 Semiconductor Types

Semiconductors can be classified in several ways. A first distinction is drawn between **intrinsic** and **extrinsic** semiconductors. Intrinsic semiconductors are characterized by a relatively small band gap and the absence of doping. At absolute zero (T = 0 K), the Fermi level corresponds to an occupation probability of 1/2. Extrinsic semiconductors, by contrast, contain intentionally introduced impurities that modify their electronic properties. They can be further categorized as follows:

- **n-type semiconductors (donors):** in these materials, doping introduces additional electrons, shifting the Fermi level E_F upward and bringing it closer to the conduction band;

- **p-type semiconductors (acceptors):** in this case, doping introduces holes, shifting E_F downward toward the valence band.

A second distinction separates semiconductors into **degenerate** and **non-degenerate** materials:

- **degenerate semiconductors** possess a doping concentration sufficiently high to shift E_F into the conduction band (for electron-doped materials) or into the valence band (for hole-doped materials), as illustrated in Fig. 5.6. In this configuration, the material behaves like a metal, since Fermi level lies within the energy bands. For degenerate semiconductors, the Fermi level essentially coincides with the Fermi energy, which sharply divides the occupied from the unoccupied electronic states. Their behavior is metallic, and their charge carrier distribution follows the relation:

$$n = \frac{k_F^3}{3\pi^2},$$

where k_f is the Fermi wavevector;

- **non-degenerate semiconductors** are materials in which the Fermi level remains within the band gap. These include intrinsic semiconductors as well as moderately doped extrinsic ones. Experimentally, a non-degenerate semiconductor is identified by the condition:

$$E_F \begin{cases} < E_C - 3k_BT \\ > E_V + 3k_BT, \end{cases}$$

where E_C and E_V denote the conduction and valence band edges, respectively, k_B is the Boltzmann constant, and T is the temperature.

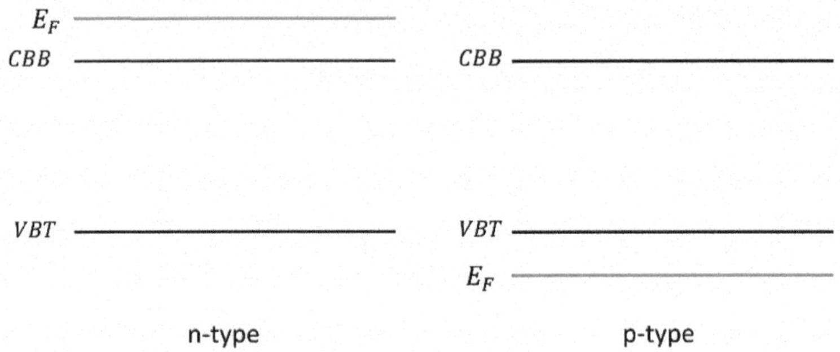

Fig. 5.6 Band structure of n-type and p-type degenerate semiconductor materials. In n-type semiconductors, E_F intersects the conduction band due to the addition of electrons, whereas in p-type semiconductors, E_F intersects the valence band due to the addition of holes

5.4 The Effective Mass

An essential concept closely related to the density of states is the **effective mass**. It is well established that an electron in a periodic potential can be regarded as "semi-free": its plane-wave function is modulated by a periodic function sharing the periodicity of the crystal lattice. At the edges of the Brillouin zones, the energy band is deformed by the periodic potential, and near the Bragg planes the normal component of the group velocity vanishes. Here the degeneracy (corresponding to wavevectors $\vec{k}$ and $\vec{k} + \vec{G}$) splits, producing two distinct energy branches and opening a band gap. At these points a **classical inversion** occurs: the electron effectively "reflects" from the Bragg plane, consistent with the von Laue diffraction condition. Because these are inversion points, the electron velocity v vanishes. The velocity, defined as the **group velocity** (remembering that the phase velocity is defined as ω/k) is given by:

$$v = \frac{\partial \omega}{\partial k},$$

where:

$$\omega = \frac{\varepsilon_n}{\hbar}.$$

Then:

$$v = \frac{1}{\hbar}\frac{\partial \varepsilon_n(k)}{\partial k}.$$

At points where the derivative of the energy eigenvalue vanishes, the velocity is zero. However, away from these points the velocity remains well defined through the slope of the dispersion relation. This leads naturally to the introduction the **effective mass**. Although $\hbar k$ is not a momentum, one may introduce a quantity with dimensions of mass, denoted m^*, which allows $\hbar k$ to be expressed as a momentum in the form:

$$\hbar k = m^* v.$$

This effective mass incorporates the effects of the band structure and allows one to treat the electron "as if" it were free, though with a mass different from the actual electron mass m_e. Using the expression for the group velocity, we write:

$$\hbar k = m^* \frac{1}{\hbar}\frac{\partial \varepsilon_n(k)}{\partial k}.$$

Differentiating both sides with respect to t gives:

$$\hbar \frac{dk}{dt} = m^* \frac{1}{\hbar}\frac{\partial^2 \varepsilon_n(k)}{\partial k^2}\frac{dk}{dt}.$$

Assuming k depends only on time, the factors dk/dt cancel, yielding:

$$\hbar = m^* \frac{1}{\hbar} \frac{\partial^2 \varepsilon_n(k)}{\partial k^2}.$$

Rearranging provides the definition of the effective mass:

$$m^* = \frac{\hbar^2}{\frac{\partial^2 \varepsilon_n(k)}{\partial k^2}}.$$

The curvature of the dispersion relation (i.e., its concavity or convexity) thus determines the inverse of the electron's inertia. In three dimensions, where the energy eigenvalues $\varepsilon_n(\vec{k})$ depend on the components k_i ($i = 1, 2, 3$), the effective mass generalizes to a second-rank tensor:

$$m_{ij} = \frac{\hbar^2}{\frac{d^2 \varepsilon_n(\vec{k})}{dk_i dk_j}}.$$

To examine the behavior near the conduction-band minimum, we expand the energy eigenvalue:

$$\varepsilon_n(\vec{k}) = E_C + \sum_i \frac{\partial \varepsilon_n}{dk_i} k_i + \frac{1}{2} \sum_{i,j} \frac{\partial^2 \varepsilon_n}{\partial k_i \partial k_j} k_i k_j = E_C + \frac{\hbar^2}{2} \sum_{i,j} m_{ij}$$

and at the conduction band minimum, the linear term vanishes, leading to:

$$\varepsilon_n(\vec{k}) \approx E_C + \frac{\hbar^2 k_x^2}{2m_x^*} + \frac{\hbar^2 k_y^2}{2m_y^*} + \frac{\hbar^2 k_z^2}{2m_z^*}.$$

By averaging the principal components, one obtains the **effective electron mass**:

$$\frac{m_x^* + m_y^* + m_z^*}{3} \cong m_n.$$

The energy eigenvalue then takes the isotropic form:

$$\varepsilon_n(\vec{k}) = E_C + \frac{\hbar^2 k^2}{2m_n}.$$

which explicitly defines the effective mass m_n. An analogous treatment applies to the valence-band maximum, as will be discussed in the following sections.

5.5 Non-degenerate Semiconductors

Let us summarize some key concepts regarding non-degenerate semiconductors. The **Fermi level** is the electrochemical potential of electrons ($E_F = \overline{\mu}$), and their occupation follows the Fermi–Dirac distribution:

$$f_n(E) = \frac{1}{1 + e^{\frac{E-\overline{\mu}}{K_B T}}}.$$

Assuming the valence band is completely occupied, consider the situation in which an electron is excited into the conduction band. As illustrated schematically in Fig. 5.7, promoting an electron to the conduction band leaves behind a vacancy (namely, a **hole**) in the valence band. The remaining electrons in the valence band then shift collectively in one direction (e.g., to the right), whereas the holes, which can be treated as positive charge carriers, appear to move in the opposite direction (e.g., to the left).

Since the hole distribution $f_p(E)$ follows from the complementarity of electronic states, we use:

$$f_n(E) + f_p(E) = 1$$

to derive it:

$$f_p(E) = 1 - f_n(E) = 1 - \frac{1}{1 + e^{\frac{E-E_F}{K_B T}}} = \frac{1 + e^{\frac{E-E_F}{K_B T}} - 1}{1 + e^{\frac{E-E_F}{K_B T}}} = \frac{1}{1 + e^{\frac{E_F-E}{K_B T}}}.$$

Another key feature of non-degenerate semiconductors concerns the band gap type, which may be direct or indirect. The conduction-band minimum and the valence-band maximum correspond to specific crystal momenta $\vec{k}$ in the Brillouin

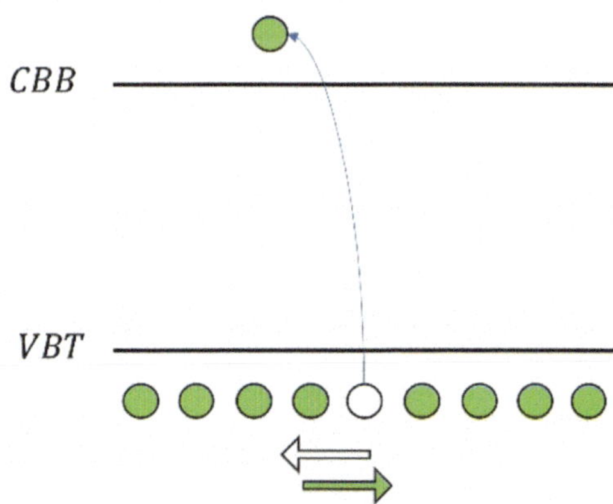

Fig. 5.7 Electronic current and hole current, whose direction is identified by the green and the white arrow, respectively

zone. When these two $\vec{k}$-values differ, the semiconductor exhibits an **indirect band gap**; if they coincide, the material has a **direct band gap**.

For a direct-gap semiconductor, an electron at $T = 0$ K may undergo a direct transition from the valence band to the conduction band via optical absorption (i.e., by absorbing a **photon**), conserving crystal momentum and potentially emitting a photon upon recombination. Conversely, for an **indirect gap**, such a direct transition is forbidden at $T = 0$ K because it must first pass through an intermediate state, transferring momentum to the crystal lattice. This process requires the participation of a **phonon**, with the phonon momentum matching the difference between the electron and hole momenta. This phonon-assisted mechanism ensures conservation of both momentum and energy. A schematic representation is provided in Fig. 5.8.

Building on this foundation, let us derive the expression for $\varepsilon_n(k)$ in non-degenerate n-type and p-type semiconductors. At Bragg planes, the energy dispersion reaches a stationary point, implying:

$$\frac{d\varepsilon_n(k)}{dk} = 0.$$

Expanding the energy near the conduction-band edge and truncating at second order (parabolic approximation) yields:

$$\varepsilon_n(k) = E_C + \frac{d\varepsilon_n}{dk}k + \frac{d^2\varepsilon_n}{dk^2}k^2 = E_C + \frac{d^2\varepsilon_n}{dk^2}k^2,$$

where the linear term vanishes at the stationary point.

Multiplying and dividing the second term by $\hbar^2 k^2$ gives:

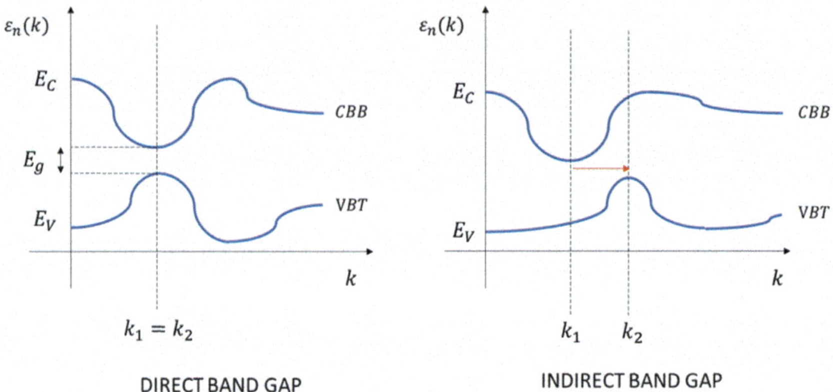

Fig. 5.8 Direct and indirect band gap. CBB and VBT indicate the conduction band bottom and the valence band top, respectively. In the direct gap the maximum of the VBT and the minimum of the CBB are at the same $k(k_1 = k_2)$; in the indirect band gap, the maximum of the VBT and the minimum of the CBB are at different k $(k_1 \neq k_2)$

$$\varepsilon_n(k) = E_C + \frac{\hbar^2 k^2}{\frac{2\hbar^2}{\frac{d^2\varepsilon_n(k)}{dk^2}}} = E_C + \frac{\hbar^2 k^2}{2m_n},$$

using the definition of the **effective mass** in one dimension already found in Sect. 5.4:

$$m_n = \frac{\hbar^2}{\frac{d^2\varepsilon_n(k)}{dk^2}}.$$

This represents dynamic mass of an electron responding to external forces within the periodic potential and is distinct from the free-electron mass. Moreover, we know that, for higher dimensions, the effective mass is represented by a tensor:

$$m_{ij} = \frac{\hbar^2}{\frac{d^2\varepsilon_n(k)}{dk_i dk_j}},$$

where $i, j = 1, 2, 3$ in a three-dimensional system. This tensor is particularly useful for materials with **anisotropic effective masses** along different crystallographic directions. The acceleration a_i experienced by an electron in each direction i due to a force $\vec{F}$ is given by:

$$a_i = \sum_j \frac{F_j}{m_{ij}}.$$

It is always possible to diagonalize the effective mass tensor, yielding three principal components corresponding to effective masses m_1, m_2, and m_3. The effective mass can then be approximated to the first order as:

$$m_n \sim \sqrt[3]{m_1 m_2 m_3},$$

which corresponds to the **density-of-states effective mass**. This value encapsulates key information about the conduction and valence bands. With this approximation, the energy near the band edges becomes:

$$\varepsilon_n(k) \begin{cases} E_C + \frac{\hbar^2 k^2}{2m_n}, \text{ for electrons} \\ \\ E_V - \frac{\hbar^2 k^2}{2m_p}, \text{ for holes,} \end{cases}$$

where m_p is the hole effective mass. The second equation is derived by expanding the energy near the maximum of the valence band. This formulation applies to cases with particular symmetry, where the problem can be analyzed from both perspectives.

With these concepts, we can now proceed in the calculation of the **density of states**. Let us begin by recalling the expression for the electron density in a generic $\vec{k}$-space volume (derived in Sect. 5.1):

$$n = \frac{2}{(2\pi)^3} V_{\vec{k}-\text{space}}.$$

For non-degenerate states, the Fermi-Dirac distribution is given by:

$$f_n(\varepsilon) = \frac{1}{1 + \frac{\varepsilon - E_F}{K_B T}}.$$

Integrating over the first Brillouin zone gives:

$$n = \frac{2}{(2\pi)^3} \int_{1\text{BZ}} \frac{1}{1 + e^{\frac{\varepsilon_n(\vec{k}) - E_F}{K_B T}}} d^3k.$$

Substituting the parabolic form for $\varepsilon_n(\vec{k})$ and using $d^3k = 4\pi k^2 dk$, extending the integral from 0 to ∞ (all relevant information are contained in the first Brillouin zone), we get:

$$n = \frac{2}{(2\pi)^3} \int_0^{\infty} \frac{1}{1 + e^{\frac{E_C + \frac{\hbar^2 k^2}{2m_n} - E_F}{K_B T}}} 4\pi k^2 dk.$$

Introducing new variables:

$$\begin{cases} -\eta = \frac{E_C - E_F}{K_B T} \\ x = \frac{\hbar^2 k^2}{2m_n K_B T}. \end{cases},$$

then:

$$\begin{cases} k = \sqrt{\frac{2m_n K_B T}{\hbar^2}} x^{\frac{1}{2}} \\ dk = \frac{1}{2} \sqrt{\frac{2m_n K_B T}{\hbar^2}} x^{-\frac{1}{2}} dx. \end{cases}$$

Substituting these expressions, the integral becomes:

$$n = \frac{1}{2\pi^2} \left(\frac{2m_n K_B T}{\hbar^2} \right)^{\frac{3}{2}} \int_0^{\infty} \frac{x^{\frac{1}{2}}}{1 + e^{x-\eta}} dx$$

and this simplifies further to:

$$n = 2 \left(\frac{m_n K_B T}{2\pi \hbar^2} \right)^{\frac{3}{2}} \frac{2}{\sqrt{\pi}} \int_0^{\infty} \frac{x^{\frac{1}{2}}}{1 + e^{x-\eta}} dx.$$

Let us define:

$$n = N_C F(\eta),$$

where N_C is the **effective density of states** in the conduction band:

$$N_C = 2\left(\frac{m_n K_B T}{2\pi \hbar^2}\right)^{\frac{3}{2}},$$

and $F(\eta)$ is the **Fermi integral**:

$$F(\eta) = \frac{2}{\sqrt{\pi}} \int_0^\infty \frac{x^{\frac{1}{2}}}{1 + e^{x-\eta}} dx.$$

To ensure the semiconductor remains non-degenerate, the condition is:

$$\frac{E_C - E_F}{K_B T} \gg 1,$$

so

$$\eta \ll -1.$$

In this regime, and by multiplying and dividing the Fermi integral for e^η, the integral simplifies as:

$$F(\eta) = \frac{2}{\sqrt{\pi}} e^\eta \int_0^\infty \frac{x^{\frac{1}{2}}}{e^\eta + e^x} dx \approx \frac{2}{\sqrt{\pi}} e^\eta \int_0^\infty x^{\frac{1}{2}} e^{-x} dx.$$

The integral equals the **Gamma function** $\Gamma(3/2)$:

$$\Gamma\left(\frac{3}{2}\right) = \int_0^\infty x^{\frac{1}{2}} e^{-x} dx = \frac{\sqrt{\pi}}{2}$$

therefore:

$$F(\eta) \approx \frac{2}{\sqrt{\pi}} e^\eta \frac{\sqrt{\pi}}{2} = e^\eta.$$

Using this result, the electron density simplifies to:

$$n = N_C e^{\frac{-(E_C - E_F)}{K_B T}}.$$

For symmetry reason, by an analogous calculation for the valence band, the hole density p becomes:

$$p = N_V e^{\frac{E_V - E_F}{K_B T}},$$

where:

$$N_V = 2\left(\frac{m_p K_B T}{2\pi \hbar^2}\right)^{\frac{3}{2}}$$

is the effective density of states in the valence band.

Both expressions feature an exponential temperature dependence governed by the denominator $k_B T$, while N_C and N_V scale as $T^{3/2}$. These two relations form a cornerstone of semiconductor physics.

5.6 Intrinsic Semiconductors

Let us now calculate the position of the Fermi level for **intrinsic semiconductors**. An intrinsic semiconductor is a chemically pure (undoped) material in which the charge carrier density is determined solely by the material properties, without contributions from impurities. In intrinsic semiconductors, the number of excited electrons in the conduction band equals the number of holes in the valence band.

The **mass-action law** is a fundamental relation governing carrier populations in semiconductors, which states that at a given temperature the product of the electron concentration n and the hole concentration p is constant and independent of impurity levels. Therefore, it applies to both intrinsic and extrinsic semiconductors. Using the expressions for n and p derived in Sect. 5.3, one obtains:

$$np = N_C N_V e^{\frac{-(E_C - E_V)}{K_B T}} = N_C N_V e^{\frac{-E_g}{K_B T}} = n_i^2$$

where n_i is the **intrinsic carrier concentration** and $E_g = E_C - E_V$ is the band gap. This expression shows that n_i depends only on material parameters (i.e., E_g, m_n, and m_p) and on the temperature, but not on the Fermi level. For a typical semiconductor with $E_g \approx 1.1$ eV, $N_C \sim N_V \sim 10^{19}\text{cm}^{-3}$, and at room temperature ($T = 300$ K, where $K_B T = 0.026$ eV), the intrinsic carrier concentration is:

$$n_i = \sqrt{N_C N_V} e^{\frac{-E_g}{2K_B T}} \approx 10^{11}\text{cm}^{-3}.$$

This result highlights that the intrinsic carrier density is much smaller than the effective density of states in the conduction and valence bands (N_C, N_V). Although the free carrier density is relatively small compared to the total charge in the material, it is crucial for electrical conduction.

Since $n \equiv p$ in intrinsic semiconductors, the Fermi level position can be found by equating the carrier densities in the two bands:

$$N_C e^{\frac{-(E_C - E_F)}{K_B T}} = N_V e^{\frac{E_V - E_F}{K_B T}}.$$

Rearranging terms and taking the natural logarithm yields:

$$E_F - E_C = K_B T \ln\left(\frac{N_V}{N_C}\right) + E_V - E_F,$$

and the Fermi level position is given by:

$$E_F = \frac{K_B T}{2} \ln\left(\frac{m_p}{m_n}\right)^{\frac{3}{2}} + \frac{E_V + E_C}{2}.$$

Simplifying further:

$$E_F = \frac{3}{4} K_B T \ln\left(\frac{m_p}{m_n}\right) + \frac{E_V + E_C}{2}.$$

In the limit $T \to 0$, the first term vanishes, and the Fermi level lies exactly at the midpoint of the band gap:

$$E_F = \frac{E_V + E_C}{2}.$$

At zero temperature, the Fermi surface concept is not meaningful for a semiconductor, because no states exist inside the gap. Only the Fermi level itself is defined, as illustrated schematically in Fig. 5.9.

At finite but low temperatures ($T \neq 0$ K), the Fermi level deviates slightly from midpoint of the bandgap for an intrinsic semiconductor. Using the previously derived equation for E_F, the shift is determined by the first term:

$$A = \frac{3}{4} K_B T \ln\left(\frac{m_p}{m_n}\right)$$

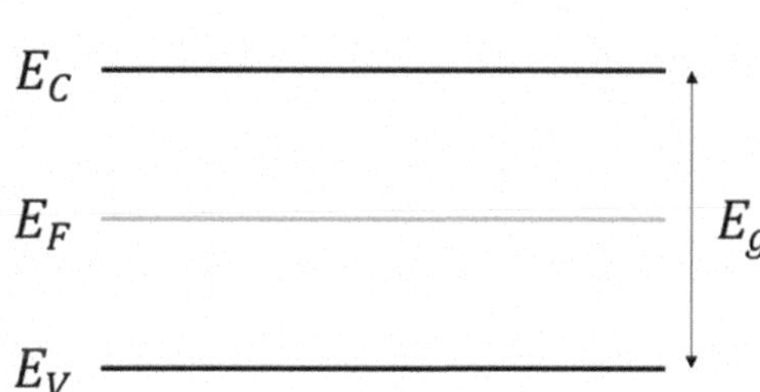

Fig. 5.9 Fermi level position in an intrinsic semiconductor for $T \to 0$ K. The Fermi level lies exactly in the midpoint of the band gap

which depends on the relative magnitudes of the electron effective mass m_n and the hole effective mass m_p. The term A can be either positive or negative, depending on the material's structural properties. In both cases, A is small at low temperatures, so E_F does not deviate significantly from its position within the bandgap (for reasonably low T, which is the range of interest).

Two possible scenarios arise:

1. **case 1:** $m_p < m_n \Rightarrow N_V < N_C \Rightarrow A < 0 \Rightarrow E_F$ shifts slightly downward;
2. **case 2:** $m_p > m_n \Rightarrow N_V > N_C \Rightarrow A > 0 \Rightarrow E_F$ shifts slightly upward.

In both situations, the displacement from the middle of the gap remains small for temperatures of practical interest. Only at very high temperatures (well beyond typical operating conditions) the shift becomes appreciable.

5.7 Extrinsic Semiconductors

Extrinsic semiconductors are materials whose electrical properties have been deliberately modified by the controlled introduction of impurity atoms, **dopants**, during crystal growth. These dopants introduce additional energy levels within the band gap and thereby alter the carrier concentrations relative to the intrinsic case. As a result, the electrical conductivity can be significantly tailored.

There are two primary types of dopants:

- **donors**: atoms that "donate" mobile electrons to the host crystal. A semiconductor doped with donors is termed **n-type**, where electrons are the majority carriers;
- **acceptors**: atoms that "accept" electrons from the valence band, thereby generating mobile holes. A semiconductor doped with acceptors is termed **p-type**, with holes as the majority carriers.

Since doping breaks the electron–hole symmetry characteristic of intrinsic semiconductors, the electron and hole densities satisfy $n \neq p$. Nonetheless, the **mass-action law** remains valid:

$$np = n_i^2,$$

where n_i is the intrinsic carrier concentration.

In technologically relevant semiconductors, most dopants are **shallow**, i.e., their energy levels lie very close to the band edges: donor levels slightly below the conduction band bottom (CBB), and acceptor levels slightly above the valence band top (VBT), as shown in Fig. 5.10. Their ionization energies are therefore very small, such that at moderate temperatures electrons are readily thermally excited from donor levels to the conduction band, or from the valence band to acceptor levels (leaving holes behind). At $T = 0$ K, however, thermal energy is insufficient to ionize the dopants, and therefore carriers arising from these states are absent. This changes as the temperature increases.

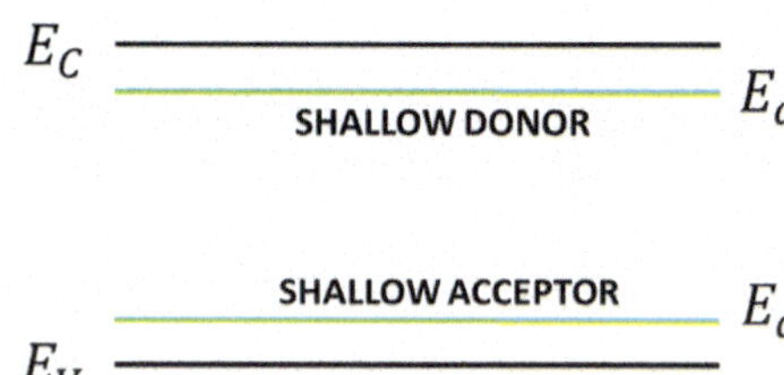

Fig. 5.10 Shallow donor (of energy E_d) and shallow acceptor (of energy E_a) states. The first can be positively ionized, the second can be negatively ionized with small activation energies

Let us consider the effect of impurities that introduce defect states within the band gap, treating them through a simplified model. These defects may be regarded as atomic-like centers, which allows the use of the same formalism applied to the **hydrogen atom**. In particular, the donor energy level E_d created by the impurity locally breaks the crystal symmetry. Conceptually, the impurity behaves like a hydrogen atom embedded in the solid: a crystalline electron is bound to a positively charged ion (H^+) rather than freely propagating through the lattice.

Under this assumption, the donor (and analogously the acceptor) energy level can be estimated as:

$$E_d = 1R_y\left(\frac{\varepsilon_0}{\varepsilon_S}\right)\left(\frac{m_n}{m_e}\right)^2 \sim 1\text{meV}$$

where $R_y = 13.6\,\text{eV}$ is the Rydberg energy, $\varepsilon_S = \varepsilon_0\varepsilon_r$ is the permittivity of the semiconductor (ε_r being the relative dielectric constant), with $\varepsilon_0/\varepsilon_S \sim 1/10$ and $(m_n/m_e)^2 \sim 1/100$.

This resulting value $E_d \sim 1\,\text{meV}$ indicates that the donor level is located very close to the CBB. Similarly, acceptor levels are located near the VBT. Despite their energetic proximity to the band edges, the spatial extent of these defect states is exceptionally large, typically encompassing about a million lattice sites. Although these defects preserve some degree of local symmetry, they nonetheless disrupt the global periodicity of the crystal. Dopant levels appear within the band gap because the symmetry-breaking nature of defects prevents them from forming Bloch states, which would otherwise be forbidden in the gap region.

Using the hydrogen atom framework, the defect radius can be expressed as:

$$r = a_B\left(\frac{\varepsilon_S}{\varepsilon_0}\right)\left(\frac{m_e}{m_n}\right) \sim 100a_B$$

where the Bohr radius is $a_B \sim 0.53$ Å, $\varepsilon_S/\varepsilon_0 \sim 10$, and $m_e/m_n \sim 10$. This confirms the significant spatial extent of the defect radius.

Let us now determine the Fermi level in extrinsic semiconductors. Experimentally, its position is described by:

$$E_F = \frac{E_C + E_d}{2} + \frac{1}{2}K_BT\ln\left(\frac{N_d}{gN_C}\right),$$

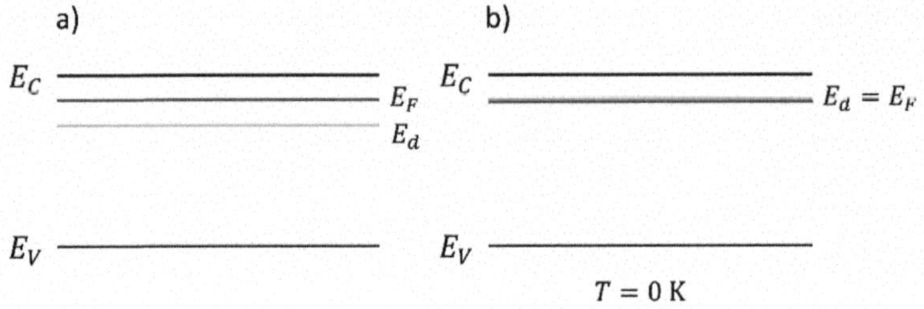

Fig. 5.11 Fermi level position in n-type extrinsic semiconductor at $T \neq 0$ K (**a**) and at $T = 0$ K (**b**)

where g is a degeneracy factor, N_d is the donor concentration, and N_C is the effective density of states in the conduction band. The donor level lies below the Fermi level, as illustrated in Fig. 5.11a.

At $T \to 0$ K, this relationship no longer holds, and the Fermi level coincides with the donor level: $E_F = E_d$, rather than $E_F = (E_C + E_d)/2$. For acceptors, the same reasoning holds, replacing E_d with E_a (acceptor energy level) and N_d with N_a (acceptor concentration), as illustrated in Fig. 5.11b.

At $T = 0$ K, the extrinsic semiconductor is non-degenerate: both the Fermi level and the donor level lie within the band gap.

Let us now see what occurs **at very high temperatures** ($T \gg 0$ K). In the n-type case, at sufficiently high temperatures all donor states become ionized. Consequently, the donor concentration N_d is approximately equal to the concentration of ionized donors N_d^+, such that:

$$N_d \cong N_d^+ \sim 10^{18}\text{cm}^{-3} \gg n_i \sim 10^{11}\ \text{cm}^{-3}.$$

Within the hydrogen atom model, nearly each ionized donor contributes one electron to the conduction band, so $n \approx N_d^+$. By applying the mass-action law $n_i^2 = np$, the hole density becomes:

$$p = \frac{n_i^2}{n} \approx \frac{10^{22}}{10^{18}} \approx 10^4 \ll n_i.$$

Thus, $p \approx 0$, confirming that holes are minority carriers in n-type materials and can be neglected for conduction. From the expression for the electron density, one obtains:

$$N_d \approx n = N_C e^{\frac{-(E_C - E_F)}{k_B T}},$$

which leads to

$$k_B T \ln\left(\frac{N_d}{N_C}\right) = E_F - E_C.$$

Therefore, the Fermi level is given by:

$$E_F = E_C + k_B T \ln\left(\frac{N_d}{N_C}\right).$$

An analogous situation occurs for p-type semiconductors, where electrons are the minority carriers ($n \approx 0$). In this case:

$$N_a \approx p = N_V e^{\frac{E_V - E_F}{k_B T}},$$

leading to

$$k_B T \ln\left(\frac{N_a}{N_V}\right) = E_V - E_F,$$

and the Fermi level is:

$$E_F = E_V - k_B T \ln\left(\frac{N_a}{N_V}\right).$$

This can be interpreted as follows: when the temperature increases, the Fermi level moves downward for n-type semiconductors and upward for p-type semiconductors. An example of extrinsic semiconductors includes the combination of silicon (Si) with dopants such as phosphorus (P) or boron (B), schematically represented in Fig. 5.12.

P belongs to Group V of the periodic table and has five valence electrons. When incorporated into the tetrahedral lattice of Si, which has four valence electrons, P readily donates electrons to the conduction band. This process generates a localized defect near the conduction band that requires only minimal amount of energy to ionize (such as thermal excitation). Conversely, B is a Group III element with three valence electrons. When added to Si, it creates a hole in the valence band by introducing a defect in the crystal structure that readily captures an electron from the lattice with minimal thermal excitation. These processes are schematically illustrated in Fig. 5.13.

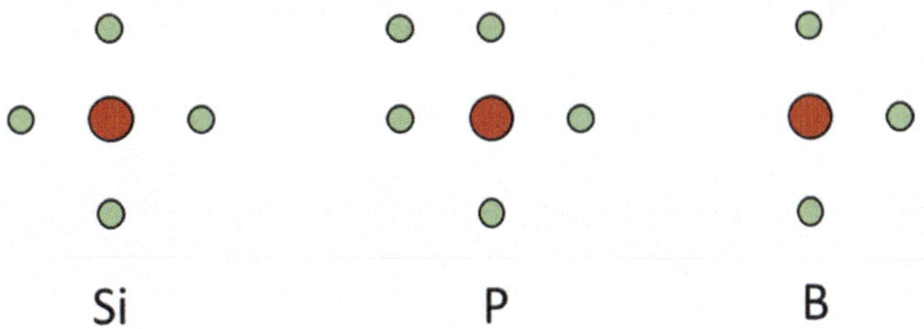

Fig. 5.12 Schematic representation of the valence electrons configuration of Si, P, and B

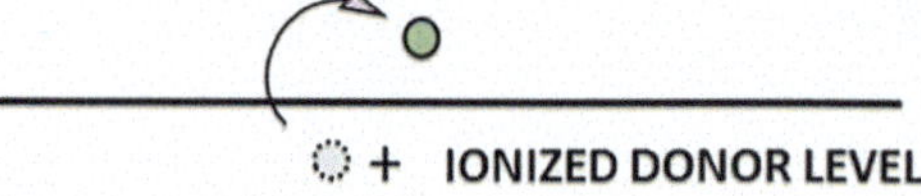

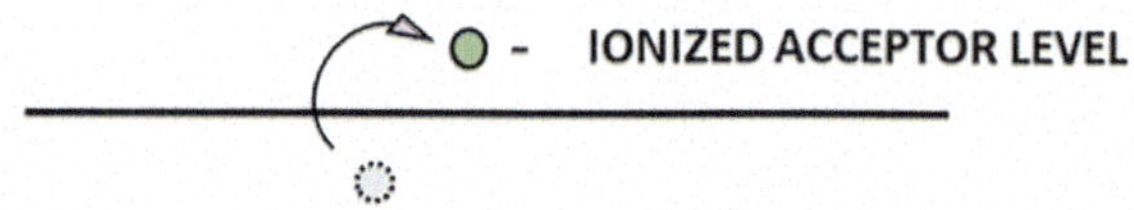

Fig. 5.13 Schematic representation of the ionization of a donor and an acceptor level with a small thermal excitation

The density of donors (N_d) and acceptors (N_a) must be maintained within the range of

$$N_d, N_a \sim 10^{16} - 10^{18}\text{cm}^{-3},$$

to avoid entering the degenerate regime. These values are far smaller than the atomic density of silicon, which is approximately 10^{23} cm^{-3}.

5.8 Band Bending

Now let us focus on the concept of **band bending**, which refers to the spatial curvature of the energy bands within a material. To illustrate this from a physical standpoint, consider a block of non-degenerate extrinsic semiconductor material (Fig. 5.14) in which a non-uniform doping profile has been introduced during fabrication.

For simplicity, consider a one-dimensional configuration in which the non-uniformity develops along the x-axis. In this scenario, the doping concentration varies spatially (for example, Si doped with a higher concentration of P atoms on the left side than on the right). This leads to a higher electron concentration n^{++} on the left and a lower concentration n on the right. Thus, more free electrons reside on the left side, giving rise to a net diffusion current from left to right.

The resulting flow of electrons leaves behind positively charged donor ions on the left and accumulates excess electrons on the right, thereby generating a potential difference across the material. An electric field $\vec{F}$ arises which drives electrons back toward their original region. Consequently, two opposing electron motions occur simultaneously: **diffusion** (from left to right) and **drift** (from right to left) due to the influence of the electric field. At equilibrium, these two processes balance each other, resulting in zero net current through the material. This equilibrium is achieved under thermodynamic conditions where temperature T is constant.

Beyond the balance of drift and diffusion, the system must also satisfy **electrochemical equilibrium**, meaning that the Fermi level must remain constant

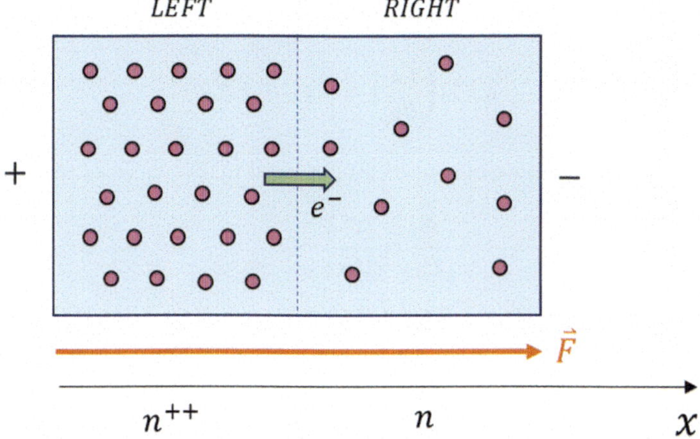

Fig. 5.14 Schematic representation of a non-degenerate extrinsic semiconductor material with two different doping concentrations on the left- and right-hand side

throughout the material:

$$\overline{\mu}^e = \text{const}.$$

The Fermi level, interpreted as the electrochemical potential, can be expressed for the left and right sides of the material as follows:

- on the **left**:

$$\overline{\mu}^e_L = \mu_0 + KT \ln\left(\frac{P_L}{P_0}\right) - e\phi_L;$$

- on the **right**:

$$\overline{\mu}^e_R = \mu_0 + KT \ln\left(\frac{P_R}{P_0}\right) - e\phi_R.$$

At equilibrium, the Fermi level must remain constant throughout the material, meaning:

$$\overline{\mu}^e_L = \overline{\mu}^e_R.$$

Substituting the expressions for the electrochemical potentials yields:

$$\mu_0 + KTl \ln\left(\frac{P_L}{P_0}\right) - e\phi_L = \mu_0 + KT \ln\left(\frac{P_R}{P_0}\right) - e\phi_R,$$

and simplifying:

$$KT \ln\left(\frac{P_L}{P_R}\right) = e(\phi_L - \phi_R).$$

Since the partial pressure of a gas is proportional to its concentration ($P \propto [C]$):

$$\frac{P_L}{P_R} \propto \frac{[C_L]}{[C_R]} > 1,$$

that implies:

$$e(\phi_L - \phi_R) \geq 0.$$

This result indicates that a spatial difference in concentration between the two sides of the material necessarily induces an electrostatic potential difference. To maintain a constant Fermi level, there must be an equilibrium between the chemical and electrical contributions. Although a voltmeter placed across the two sides would not register a potential difference, there exists an intrinsic electrostatic potential difference reflected by the constant Fermi level. The net current is ultimately zero, and:

$$\phi_L > \phi_R.$$

From an energetic point of view, the carrier density can be expressed as:

$$n(x) = N_C e^{\frac{E_F - E_C(x)}{K_B T}}.$$

Since E_F and N_C are constants, the spatial dependence of carrier density $n(x)$ implies that the conduction band minimum must also vary with position, i.e., $E_C = E_C(x)$. This semiclassical treatment assumes that $E_C(x)$ varies smoothly on a macroscopic scale and does not modify the underlying crystalline band structure. Here, $E_C(x)$ is perturbed by the spatially dependent potential, which is spread over finite distances and produces fields much smaller than the crystalline field. So, we are not expanding the energy eigenvalue as a function of x (that has already been expanded as a function of k) but we are performing a perturbation.

Since $n(x)$ depends on x, its spatial derivative is non-zero:

$$\frac{\partial n(x)}{\partial x} \neq 0.$$

This implies:

$$\frac{\partial E_C(x)}{\partial x} \neq 0$$

This phenomenon is referred to as **band bending**, where the conduction band is no longer flat but acquires curvature. Energetically, the conduction band minimum relates to the potential as:

$$E_C(x) = -q\phi(x),$$

where q is the elementary charge, and $\phi(x)$ is the electrostatic potential. Additionally, knowing that:

$$\vec{\nabla}\phi = -\vec{F},$$

in one dimension, this simplifies to:

$$\frac{\partial \phi(x)}{\partial x} = -F(x),$$

and hence:

$$\frac{\partial E_C(x)}{\partial x} = qF(x).$$

Thus, the energy bands must bend in the presence of the built-in field $\vec{F}$. The same argument generalizes to p-n junctions (where one side is doped with n-type material and the other with p-type material) discussed in detail in Sect. 6.1. In Fig. 5.15 a schematic representation of the built-in field and of the conduction band minimum variation in energy along the x-direction is reported.

This phenomenon can also be approached through current densities. The electron current density of n-type carriers $\vec{j}_n$ consists of both drift and diffusion contributions:

$$\vec{j}_n = -nq\vec{v}_d + qD_n\nabla n,$$

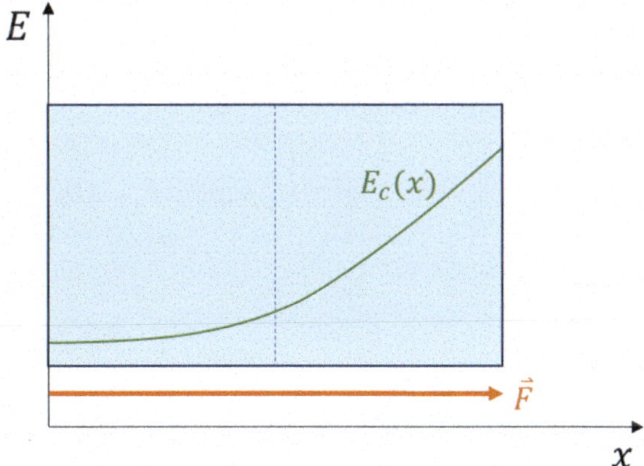

Fig. 5.15 Band-bending slope. $E_C(x)$ is the conduction band minimum and $\vec{F}$ the built-in field

where $\vec{v}_d$ is the drift velocity (induced by the electric field, which drives electrons to the left), D_n is the diffusion coefficient, and ∇n is the of carrier density gradient (which drives electrons to the right). Being:

$$\vec{v}_d = -\mu_n \vec{F},$$

where μ_n is the mobility of n-type carriers, and noting that if $\vec{F}$ is directed towards the right, the drift velocity $\vec{v}_d$ is directed towards the left, the current density can be rewritten as:

$$\vec{j}_n = nq\mu_n \vec{F} + qD_n \nabla n.$$

By assuming the derivative is taken only in the x-direction, this becomes:

$$j_n = nq\mu_n F + qD_n \frac{\partial n}{\partial x},$$

where:

$$\frac{\partial n}{\partial x} = \frac{\partial}{\partial x} N_C e^{\frac{E_F - E_C(x)}{K_B T}} = -\frac{n}{K_B T}\frac{\partial E_C(x)}{\partial x} = -\frac{nq}{K_B T} F(x).$$

Substituting this expression for $\partial n/\partial x$ into the current density equation gives:

$$j_n = nq\mu_n F(x) - qD_n \frac{nq}{K_B T} F(x).$$

At equilibrium, $j_n = 0$, yielding the **Einstein relation**:

$$\frac{D_n}{\mu_n} = \frac{K_B T}{q}.$$

This means that, for a fixed temperature, the ratio between the diffusion coefficient and the mobility remains constant. A similar reasoning applies to p-type carriers, allowing the introduction of the **quasi-Fermi level**. The hole current density in one-dimension is:

$$j_p = pq\mu_p F - qD_p \frac{\partial p}{\partial x}.$$

Since:

$$p(x) = N_V e^{\frac{E_V(x) - E_F}{K_B T}},$$

Following the same procedure as before, the derivative of $p(x)$ is given by:

$$\frac{\partial p}{\partial x} = \frac{\partial}{\partial x} N_V e^{\frac{E_V(x)-E_F}{K_B T}}.$$

When band bending occurs, all energy bands curve simultaneously, including the valence band edge, as shown in Fig. 5.16. Consequently, all energy levels follow the same slope profile, where:

$$\frac{\partial E_C(x)}{\partial x} = qF(x)$$

and

$$\frac{\partial E_V(x)}{\partial x} = qF(x).$$

Computing the derivative $\partial p/\partial x$, we obtain:

$$\frac{\partial p}{\partial x} = \frac{N_V}{K_B T}\frac{\partial E_V}{\partial x} e^{\frac{E_V-E_F}{K_B T}}$$

and substituting $\partial E_V/\partial x = qF$:

$$\frac{\partial p}{\partial x} = \frac{N_V}{K_B T} qF e^{\frac{E_V-E_F}{K_B T}} = \frac{q}{K_B T} Fp$$

Inserting this into the expression for the current density j_p, we obtain:

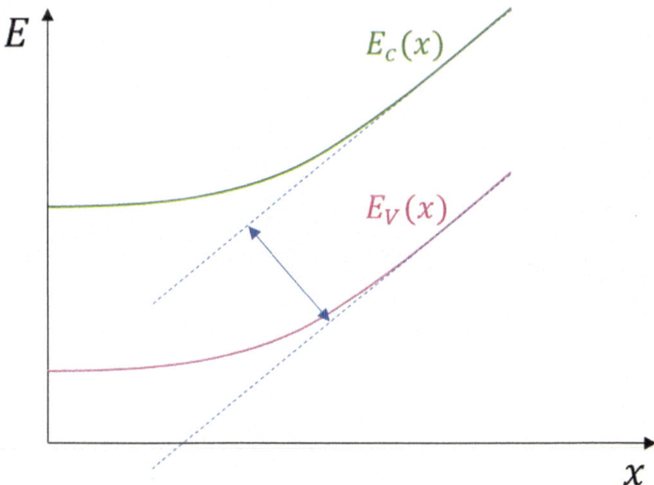

Fig. 5.16 Simultaneous band-banding for all the energy bands in the solid. $E_C(x)$ and $E_V(x)$ have the same slope (as shown by blue dashed lines)

$$j_p = pq\mu_p F - \frac{q^2 D_P F p}{K_B T}.$$

At equilibrium ($j_p = 0$):

$$\mu_p = \frac{qD_P}{K_B T}$$

and the same **Einstein relation** holds also for holes:

$$\frac{K_B T}{q} = \frac{D_P}{\mu_p}.$$

This result, combined with the corresponding relation for electrons ($D_n/\mu_n = K_B T$), shows that:

$$\frac{D_P}{\mu_p} = \frac{D_n}{\mu_n}.$$

Although derived under equilibrium, this relation between constants (D_n and D_P) at a fixed temperature T also applies in non-equilibrium regimes, enabling the definition of **quasi-Fermi level**, which is particularly useful for describing non-equilibrium conditions in gas sensors working under external bias or illumination.

In fact, when an external bias (voltage) is applied to the system, the Fermi level is no longer spatially constant. Local thermal equilibrium is preserved, while global electrochemical equilibrium is broken by the potential difference imposed by the external generator. As a consequence, a non-zero current flows through the junction or semiconductor device.

To formally address the non-equilibrium condition, the Fermi level is replaced by two quasi-Fermi levels: $\boldsymbol{E_{Fn}}$ for electrons (n-side) and $\boldsymbol{E_{Fp}}$ for holes (p-side). The electron density, previously written as:

$$n = N_C e^{\frac{E_F - E_C}{k_B T}}$$

is now expressed using the quasi-Fermi level $E_{Fn}(x)$:

$$n(x) = N_C e^{\frac{E_{Fn}(x) - E_C(x)}{k_B T}}.$$

This formulation allows a non-equilibrium situation to be treated locally as if it were in equilibrium. Unlike the Fermi level at equilibrium (which is flat throughout the material) the quasi-Fermi level varies spatially and therefore acquires a finite slope. Similarly, the hole concentration is expressed through the quasi-Fermi level for holes:

$$p(x) = N_V e^{\frac{E_V(x) - E_{Fp}(x)}{k_B T}}.$$

At equilibrium ($V_{\text{BIAS}} = 0$), the mass-action law $np = n_i^2$ holds, where:

$$n_i^2 = N_C N_V e^{\frac{-E_g}{k_B T}}$$

with $E_g = E_C - E_V$ as the energy gap. Under external bias, however, the mass-action law no longer applies. Instead, the product np becomes:

$$np = N_C N_V e^{\frac{E_V(x)-E_C(x)}{K_B T}} e^{\frac{E_{Fn}(x)-E_{Fp}(x)}{K_B T}}.$$

The first term corresponds to $n_i^2 = N_C N_V e^{\frac{-E_g}{k_B T}}$, while the second reflects the effect of the external bias. Since:

$$E_{Fn}(x) - E_{Fp}(x) = qV_{\text{BIAS}},$$

it follows that:

$$np = n_i^2 e^{\frac{qV_{\text{BIAS}}}{K_B T}}.$$

Thus, increasing V_{BIAS} raises the carrier concentrations and thereby enhances conduction. When $V_{\text{BIAS}} = 0$, the two quasi-Fermi levels coincide with the equilibrium Fermi level. In some literature, the quasi-Fermi level is also referred to as the **IMREF**.

To further interpret the role of the Einstein relation in defining the quasi-Fermi level, let us consider the electron current density j_n in a one-dimensional geometry (perpendicular to the junction):

$$j_n = nq\mu_p F + qD_n \frac{\partial n}{\partial x}.$$

Because we are now in a non-equilibrium regime, the current density does not vanish. The derivative $\partial n/\partial x$ depends on both $E_C(x)$ and $E_{Fn}(x)$:

$$\frac{\partial n}{\partial x} = N_C \frac{\partial}{\partial x} e^{\frac{E_{Fn}(x)-E_C(x)}{K_B T}} = \frac{n}{k_B T}\left(\frac{\partial E_{Fn}}{\partial x} - \frac{\partial E_C}{\partial x}\right).$$

Using $\partial E_C/\partial x = qF$, this becomes:

$$\frac{\partial n}{\partial x} = \frac{n}{k_B T}\left(\frac{\partial E_{Fn}}{\partial x} - qF\right).$$

Substituting this expression back into the current density yields:

$$j_n = nq\mu_p F + qD_n \frac{n}{k_B T}\left(\frac{\partial E_{Fn}}{\partial x} - qF\right)$$

and, using the Einstein relation $\mu_n = qD_n/(k_B T)$, this simplifies to:

$$j_n = nq\mu_p F + \mu_n n\left(\frac{\partial E_{Fn}}{\partial x} - qF\right).$$

Finally:

$$j_n = \mu_n n \frac{\partial E_{Fn}}{\partial x}.$$

An analogous result is obtained for holes:

$$j_p = \mu_p p \frac{\partial E_{Fp}}{\partial x}.$$

When an external disturbance is introduced, such as an applied potential difference or an external stimulus like illumination (e.g., in photodiodes or photovoltaic devices), the quasi-Fermi levels split and acquire different slopes, generating electron and hole currents.

If $j_n = j_p = 0$, then $\partial E_{Fn}/\partial x = \partial E_{Fp}/\partial x = 0$, the quasi-Fermi levels coincide ($E_{Fn} = E_{Fp} = E_F$), restoring the Fermi level as the electrochemical potential.

The slope of the IMREF is directly linked to the magnitude of the local current density. In a generic junction between a p-type region (left) and an n-type region (right) under applied bias, the high electron concentration on the n-side implies that a relatively small slope of E_{Fn} produces a substantial electron current j_n. Conversely, in the p-type region, where holes dominate, a larger slope of E_{Fp} is required to generate an equivalent hole current j_p. In general, the slope of the quasi-Fermi levels is directly related to the electron and hole current densities, respectively, e.g., a positive slope ($\partial E_{Fn}/\partial x > 0$) corresponds to an electron current flowing from left to right. The total current in the device is given by:

$$j_{TOT} = j_n + j_p.$$

References

1. Ashcroft NW, Mermin ND (1976) Solid state physics. Holt, Rinehart and Winston, New York
2. Kittel C (2005) Introduction to solid state physics, 8th edn. Wiley, Hoboken
3. Madou MJ, Morrison SR (1997) Chemical sensing mechanisms in solid-state devices. In: Chemical sensing with solid state devices. Academic Press, pp 74–83
4. Streetman BG, Banerjee S (2015) Solid state electronic devices, 7th edn. Global edition. Pearson
5. Pierret RF (1996) Semiconductor device fundamentals, 2nd edn. Addison-Wesley, Reading, Massachusetts
6. Sakurai JJ, Napolitano J (2017) Modern quantum mechanics, 2nd edn. Cambridge University Press, Cambridge

7. Cohen-Tannoudji C, Diu B, Laloë F (1977) Quantum mechanics, Vols. I–II. Wiley, New York
8. Davies JH (1998) The physics of low-dimensional semiconductors: an introduction. Cambridge University Press, Cambridge
9. Reif F (1965) Fundamentals of statistical and thermal physics. McGraw-Hill Science/Engineering/Math, New York
10. Yu PY, Cardona M (1996) Electronic band structures. In: Fundamentals of semiconductors. Springer, Heidelberg
11. Ridley BK (2013) Quantum processes in semiconductors, 5th edn. Oxford University Press, Oxford
12. Harrison WA (1980) Solid state theory. Dover Publications, New York
13. Lundstrom M (2000) Fundamentals of carrier transport, 2nd edn. Cambridge University Press, Cambridge
14. Madelung O (2004) Semiconductors: data handbook, 3rd edn. Springer, Berlin
15. Blakemore JS (1987) Semiconductor statistics. Dover Publications, New York
16. Shockley W (1949) The theory of p–n junctions in semiconductors and p–n junction transistors. Bell Syst Tech J 28:435–489
17. Sze SM, Ng KK (2006) Physics of semiconductor devices, 3rd edn. John Wiley & Sons, Hoboken, NJ
18. Harrison P (2005) Quantum wells, wires and dots: theoretical and computational physics of semiconductor nanostructures. Wiley, New York
19. Malagù C (2025) A thermodynamic approach to the Fermi Level in Solid State Physics: an application to gas sensing. Poster presentation at IMCS 2025, Freiburg, Germany, 24 June. Part of session: Mechanisms, modeling and simulation. Available at: https://imcs2025.com/programme/

Chapter 6
The P–N Junction

Any sufficiently advanced technology is indistinguishable from magic.
Arthur C. Clarke

The study of the p-n junction is central to understanding the behavior of semiconductor devices. Its formation and equilibrium properties provide the foundation for analyzing charge transport, potential profiles, and interfacial effects that govern device operation. Key theoretical frameworks, such as simplified approximations of the space-charge region and models addressing the role of surface states, allow a clearer description of the electronic structure and carrier dynamics within these systems. Moreover, the extension of these concepts to metal-semiconductor contacts highlights the broader relevance of junction physics in practical applications, from diodes and transistors to sensors and photovoltaic devices. This Chapter aims to present a comprehensive overview of these principles, offering both conceptual and analytical tools for a deeper understanding of semiconductor junctions and their technological importance. This Chapter results from a comprehensive analysis of the relevant literature, mainly relying on the works cited below, while extending them through the authors' own interpretative framework [1–33].

6.1 What Is a P-N Junction?

To analyze the formation of a p-n junction, let us consider the situation in which two semiconductor regions with opposite doping types are brought into contact under equilibrium. A p-n junction is created by joining a p-type semiconductor to an n-type semiconductor. For instance, silicon may be doped so that one region contains phosphorus donors (n-type) while the opposite region contains boron acceptors (p-type), producing a characteristic bending of the energy bands.

To clarify this process, imagine a p-type semiconductor on the left and an n-type semiconductor on the right, initially isolated from each other, as illustrated in Fig. 6.1.

© The Author(s), under exclusive license to Springer Nature Switzerland AG 2026
C. Malagù and G. Zonta, *Gas Sensors*, https://doi.org/10.1007/978-3-032-21614-4_6

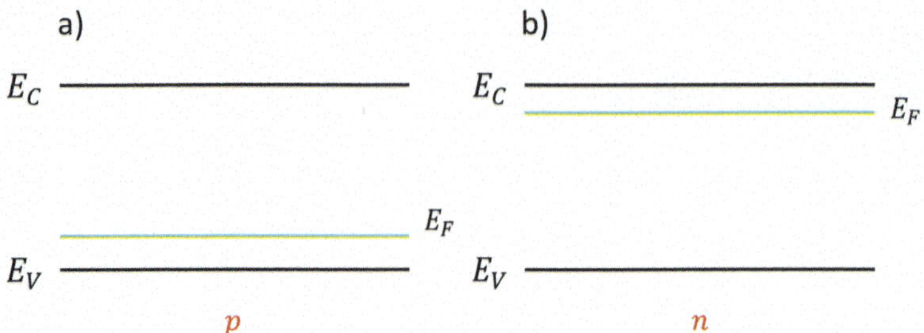

Fig. 6.1 p-type semiconductor on the left (**a**) and n-type semiconductor on the right (**b**), not in contact with each other. E_F is the Fermi level, E_C the CBB energy and E_V the VBT energy

Each material is characterized by its own Fermi level E_F, conduction-band edge E_C, and valence-band edge E_V.

When the two materials are brought into contact, the system evolves toward thermal and electrochemical equilibrium. On the n-side, electrons are in excess, while on the p-side, holes dominate. Upon contact, carriers diffuse from their respective regions of higher concentration toward regions of lower concentration: electrons migrate from the n-type side to the p-type side, and holes move in the opposite direction. Although there exists a small counterflow of carriers, the net electron flux is from n- to p-side. This diffusion process is driven by the system's tendency to maximize entropy and reach equilibrium.

As charge redistributes, the n-region acquires a positive potential relative to the p-region. In energetic terms, the energy bands on the n-side shift downward, while those on the p-side shift upward, as schematically depicted in Fig. 6.2.

At equilibrium, the Fermi level becomes uniform throughout the structure, ensuring that no net current flows across the junction and that the system has achieved thermal and electrochemical equilibrium. On the n-side, the Fermi level shifts downward because the electron concentration decreases, given that the Fermi level represents the probability of finding electrons, which is now lower than before (Fig. 6.3).

From the electrostatic viewpoint, a **built-in potential**, denoted $\boldsymbol{V_{\mathrm{BI}}}$, develops across the junction. It is positive on the n-side and negative on the p-side, consistent with the electrostatic potential profile $\Phi(x)$ shown in Fig. 6.4.

Let us now derive the expression for the built-in potential. Recall the carrier density expressions:

$$\begin{cases} n = N_C e^{\frac{E_F - E_C}{K_B T}} \\ p = N_V e^{\frac{E_V - E_F}{K_B T}} \end{cases} .$$

Consider the n-type region far from the junction ($x \to +\infty$), where the carrier concentration n is equal to the ionized donor density, N_d^+:

$$n(x \to +\infty) = N_d^+ \approx N_d$$

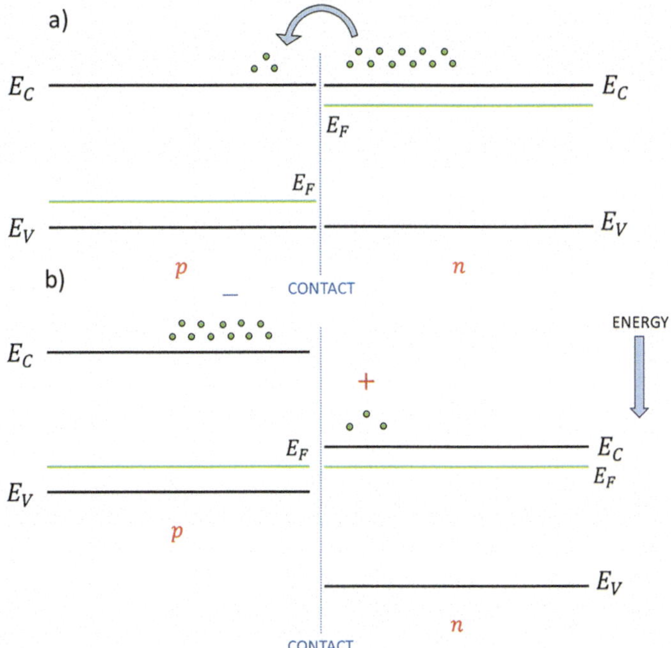

Fig. 6.2 **a** A p-type semiconductor (left) and an n-type semiconductor (right) are brought into contact. Electrons diffuse from the region of higher concentration (n-type) to the region of lower concentration (p-type), resulting in an increase in the potential and a decrease in the energy levels of the n-type material relative to the p-type material. **b** At equilibrium, the system stabilizes, and the Fermi level becomes uniform across both sides

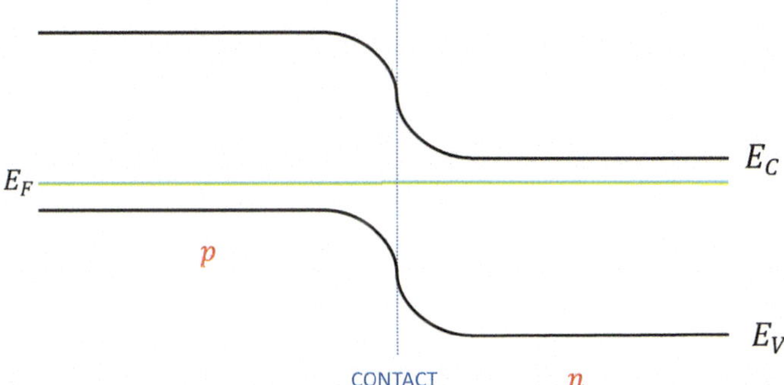

Fig. 6.3 Band bending of the energy bands: the Fermi level is now unique at both sides, and the system has achieved thermal and electrochemical equilibrium

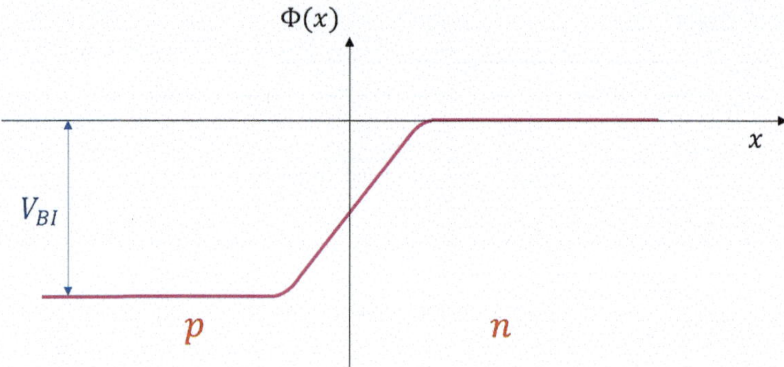

Fig. 6.4 Graph of the potential $\Phi(x)$ as a function of the position x in the material with respect to the center of the junction. The difference among the potential values in the p-type and n-type is the built-in potential V_{BI}

Under the assumption of high temperature and shallow donors, we can approximate $N_d^+ \cong N_d$. Similarly, in the p-type region far from the junction ($x \to -\infty$), the situation is mirrored:

$$p(x \to -\infty) = N_a^- \approx N_a,$$

where the ionized acceptor density approximately equals the doping level.

In the *n*-type region:

$$n(x \to +\infty) = N_d = N_C e^{\frac{E_F - E_C(x \to +\infty)}{K_B T}}.$$

Setting the reference $E_C(x \to +\infty) = 0$, we find:

$$N_C = N_d e^{-\frac{E_F}{K_B T}},$$

which holds for all values of x. Substituting this into the general expression for $n(x)$:

$$n(x) = N_d e^{-\frac{E_F}{K_B T}} e^{\frac{E_F - E_C(x)}{K_B T}} = N_d e^{-\frac{E_C(x)}{K_B T}}.$$

Since $E_C(x) = -q\Phi(\mathrm{x})$, this becomes:

$$n(x) = N_d e^{\frac{q\Phi(\mathrm{x})}{K_B T}},$$

showing that the electron concentration is directly governed by the electrostatic potential.

A similar derivation can be made for the hole concentration $p(x)$ in the p-type region, knowing that:

$$p(x \to -\infty) = N_a = N_V e^{\frac{E_V(x \to -\infty) - E_F}{K_B T}}.$$

Since $E_V(x)$=$E_C(x)-E_g$, thus $E_V(x \to -\infty) = -E_g$, and $E_C(x \to -\infty) = qV_{\mathrm{BI}}$. It follows that:

$$N_a = N_V e^{\frac{qV_{BI} - E_g - E_F}{K_B T}}.$$

Reversing this expression gives:

$$N_V = N_a e^{\frac{-qV_{BI} + E_g + E_F}{K_B T}},$$

and substituting this back into the general equation for $p(x)$:

$$p(x) = N_a e^{\frac{-qV_{BI} + E_g + E_F}{K_B T}} e^{\frac{-q\Phi(\mathrm{x}) - E_g - E_F}{K_B T}} = N_a e^{\frac{-q(V_{BI} + \Phi(\mathrm{x}))}{K_B T}}.$$

Thus, also the hole concentration depends directly on the potential. Since in the junction $\Phi(\mathrm{x}) \leq 0$ and $V_{\mathrm{BI}} + \Phi(\mathrm{x}) \geq 0$, it follows that $n(x) \leq N_d$ and $p(x) \leq N_a$, with equality achieved at the doping maximum.

Using the mass-action law:

$$n(x)p(x) = n_i{}^2 = N_d N_a e^{\frac{q\Phi(\mathrm{x})}{K_B T}} e^{\frac{-q(V_{BI} + \Phi(\mathrm{x}))}{K_B T}} = N_d N_a e^{\frac{-qV_{BI}}{K_B T}}.$$

Equating this to n_i^2 and solving for V_{BI} leads to:

$$V_{\mathrm{BI}} = \frac{K_B T}{q} \ln\left(\frac{N_d N_a}{n_i{}^2}\right).$$

This expression allows the built-in potential of any p-n junction to be calculated directly from its doping concentrations (N_d, N_a).

For typical nondegenerate doping values in silicon, the impurity concentration per lattice site is approximately 10^{-6} cm^{-3}. Considering that the atomic density in a solid is approximately 10^{23} cm^{-3}, the corresponding doping levels are typically:

$$N_d, N_a \sim 10^{17-18}\,\mathrm{cm}^{-3}.$$

The effective density of states at room temperature satisfies $N_C, N_V \sim 10^{19}$ cm^{-3}, confirming the nondegenerate condition for such doping levels. At room temperature (T~300 K) or even at higher temperatures (T~300 ° C), the intrinsic carrier concentration is approximately:

$$n_i \sim 10^{11}\,\mathrm{cm}^{-3} \ll N_d, N_a.$$

Substituting these numerical values into the expression for the built-in potential gives:

$$V_{\mathrm{BI}} \sim 1\ \mathrm{V}.$$

In nanoscale structures with characteristic junction widths on the order of 10 nm, the corresponding built-in electric fields are:

$$F_{\mathrm{BI}} \sim \frac{1V}{10^{-6} - 10^{-9}\,\mathrm{m}}.$$

For comparison, the lattice potential is of about 1 V over atomic distances of approximately 1 Å:

$$F \sim \frac{1\mathrm{V}}{1\,\mathring{\mathrm{A}}} \gg F_{\mathrm{BI}}$$

Hence, the built-in field represents only a minor perturbation and does not significantly modify the crystal band structure. Band bending (i.e., the spatial energy variation) is therefore well described by a semiclassical framework: the dispersion relation $\varepsilon_n(k)$ remains essentially unchanged, except for small second-order corrections. The junction fields are sufficiently weak that they do not appreciably perturb the underlying Hamiltonian.

6.2 Fermi Level Position in N-Type and P-Type Semiconductors

In Sect. 5.7, we discussed how defects can be modeled as hydrogen-like atoms embedded within the material, with crystalline electrons bound to a positive ion (H^+). From this model, we derived the order of magnitude of the ionization energies of shallow donors and acceptors (E_d, E_a~1 meV) and the defect radius (r~$100a_B$). We now turn our attention to the distribution of defects within the material. Let us consider the doping concentration N_d of a n-type material, which represents the maximum possible number of donor impurities. Depending on the temperature, these defects will partition into neutral (N_d^0) and positively ionized donors (N_d^+), such that:

$$N_d = N_d^0 + N_d^+.$$

An analogous expression holds for p-type materials:

$$N_a = N_a^0 + N_a^-.$$

For simplicity, we focus on the n-type case, noting that the treatment for p-type materials proceeds in the same manner.

The concentration of ionized donors N_d^+ can be written as the product of the total donor concentration and a filling probability factor that follows a Fermi-Dirac-like expression analogous to hole statistics:

$$N_d^+ = N_d \frac{1}{1 + g_d e^{\frac{E_F - E_d}{k_B T}}},$$

where g_d is the degeneracy factor, representing the number of available impurity states (typically $g_d = 2$, corresponding to the two spin orientations). At sufficiently high temperatures all donors become ionized. Here, however, we are interested in the temperature range where:

$$k_B T < E_F - E_d,$$

so that:

$$\frac{1}{1 + g_d e^{\frac{E_F - E_d}{k_B T}}} < 1,$$

meaning that not all donors are ionized. In this regime, the electron concentration may be approximated by the ionized donor density (but not to the total donor concentration):

$$n = N_C e^{\frac{E_F - E_C}{K_B T}} \approx N_d^+.$$

Using the approximation $1 + x \approx x$ for $x \gg 1$:

$$N_d \frac{1}{1 + g_d e^{\frac{E_F - E_d}{k_B T}}} \approx \frac{N_d}{g_d} e^{\frac{E_d - E_F}{k_B T}}.$$

Equating the two expressions for n:

$$N_C e^{\frac{E_F - E_C}{k_B T}} = \frac{N_d}{g_d} e^{\frac{E_d - E_F}{k_B T}}.$$

Taking the natural logarithm:

$$\ln N_C + \frac{E_F - E_C}{k_B T} = \ln \frac{N_d}{g_d} + \frac{E_d - E_F}{k_B T}$$

and rearranging to isolate E_F:.

$$E_F = \frac{k_B T}{2} \ln \frac{N_d}{N_C g_d} + \frac{E_d + E_C}{2}.$$

This expression shows that the Fermi level E_F depends on the doping concentration N_d, the effective density of states N_C, the degeneracy factor g_d, and the defect energy level E_d.

At $T = 0$ K, one might expect the Fermi level to lie at an intermediate energy between E_d and E_C. However, experimental evidences show that this is not the case and

$$E_F \neq \frac{(E_d + E_C)}{2}.$$

Instead, at very low temperatures,

$$E_F = E_d,$$

indicating that the donor activation energy aligns with the Fermi level. A similar situation occurs in p-type materials, where at $T = 0$ K the Fermi level aligns with the acceptor energy level:

$$E_F = E_a.$$

These scenarios are illustrated in Fig. 6.5.

This behavior is attributed to compensation effects (donor states present in p-type materials and acceptor states in n-type materials). In a n-type semiconductor, for

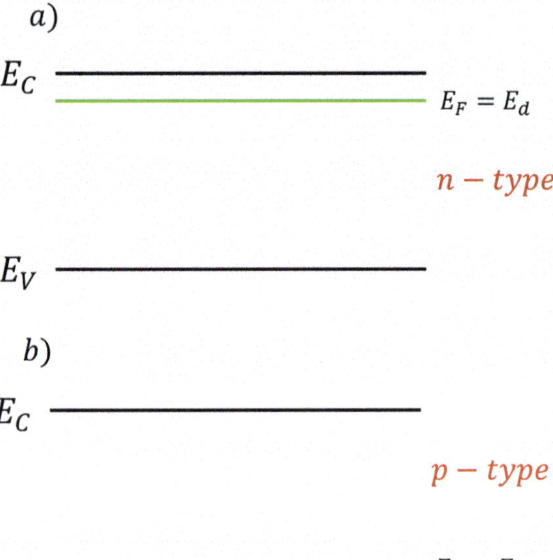

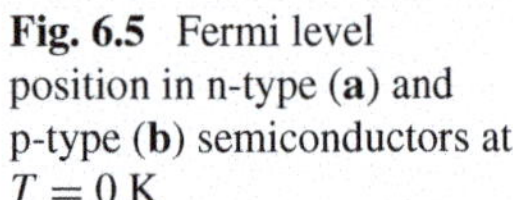

Fig. 6.5 Fermi level position in n-type (**a**) and p-type (**b**) semiconductors at $T = 0$ K

example, the Fermi level at $T = 0$ K shifts from the midpoint between the conduction band and the donor level to coincide with $E_F = E_d$.

Let us now examine the case of higher temperatures ($T \gg 0$ K). Again, considering n-type materials, under the approximation that all donors are ionized ($n \approx N_d^+ = N_d$) and neglecting the contribution of non-ionized donors in the bulk, one may write:

$$n \approx N_C e^{\frac{E_F - E_C}{k_B T}}.$$

From this, the position of the Fermi level can be determined as:

$$E_F = E_C + K_B T \ln\left(\frac{N_d}{N_C}\right).$$

Similarly, for p-type materials, assuming $p = N_a^- \approx N_a = N_V \exp\left(E_V - E_F/k_B T\right)$, the Fermi level becomes:

$$E_F = E_C - K_B T \ln\left(\frac{N_a}{N_V}\right).$$

Given that $N_d/N_C \ll 1$ and $N_a/N_V \ll 1$, it follows that that in n-type semiconductors the Fermi level decreases as temperature increases, whereas in p-type semiconductors the Fermi level moves upward with increasing temperature.

6.3 Depletion Approximation (DA)

Let us consider a **p-n junction** with the origin located at the contact interface ($x = 0$), the n-type semiconductor in the region $x \geq 0$, and the p-type semiconductor in $x \leq 0$. We assume that donors in the n-type side are fully ionized (having donated their electrons) and that acceptors in the p-type side are also ionized (having accepted electrons, thus leaving holes). Under these conditions, the **charge density** in the two regions can be written as a function of the position x as:

$$\rho(x) = q(N_d - n(x) + p_n), \quad \text{for } x \geq 0$$

$$\rho(x) = q\left(-N_a + p(x) - n_p\right), \quad \text{for x} \leq 0,$$

where q is the elementary charge (e), p_n and n_p denote the minority carrier concentrations (holes in the n-type region *n*-type and electrons in the *p*-type region, respectively). Under equilibrium, minority carriers are negligible compared with majority carriers, except when current transport (e.g., in transistors) is explicitly considered. Thus, charge densities reduce to:

$$\rho(x) = q(N_d - n(x)),$$

$$\rho(x) = q(-N_a + p(x)).$$

Expressing the carrier concentrations in terms of the local electrostatic potential $\Phi(x)$ and the built-in potential $V_{\rm BI}$, the charge densities become:

$$\rho(x) = q\left(-N_a + N_a e^{\frac{-q(V_{BI}+\Phi(x))}{k_B T}}\right)$$

$$\rho(x) = q\left(N_d - N_d e^{\frac{q\Phi(x)}{k_B T}}\right),$$

Because the charge density must vary continuously across the junction, its spatial profile can be obtained numerically (see Fig. 6.6).

Let us now consider the regime in which the electrostatic potential satisfies

$$V \gg k_B T/q,$$

so that the exponential terms become negligible in the space-charge region. Under this condition, the continuous charge distributions effectively become discontinuous. Far from the junction on the n-side ($x \to +\infty$), the potential is zero by convention:

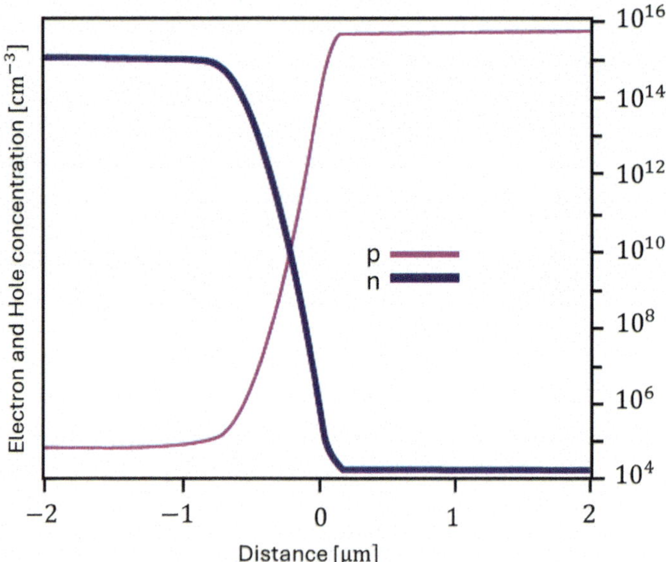

Fig. 6.6 Electron and hole concentrations (n and p) expressed in cm^{-3} as a function of the distance x from the center of the junction ($x = 0$), expressed in μm. The n-type material on the left, while the p-type material on the right

$\Phi(+\infty) = 0$. Thus:

$$\rho(x) = q(N_d - N_d) = 0.$$

A similar scenario occurs in the p-side region, where $\Phi(-\infty) = -V$ (with $V = V_{BI}$), leading to:

$$\rho(x) = q(-N_a + N_a) = 0.$$

This implies that the charge due to ionized donors (or acceptors) balances the charge due to free electrons (or holes) in the bulk regions.

Next, we focus on the **space charge region**, where the built-in electric field depletes the mobile charge carriers. Two specific points can be identified within this region: $x = x_n$ in the n-type material and $x = -x_p$ in the p-type material.

Thus, in the n-type depletion region ($0 < x < x_n$), the potential can be approximated as Φ(x)~V/2. Being $V \gg k_B T/q$, and $\Phi(x) < 0$, the charge density becomes:

$$\rho(x) \sim q\left(N_d - N_d e^{-\infty}\right) \sim qN_{d.}$$

A similar relationship holds in the space charge region of the p-type material ($-x_p < x < 0$):

$$\rho(x) = q\left(-N_a + N_a e^{-\infty}\right) \sim -qN_{a}.$$

Within the **Depletion Approximation (DA)**, in the region $-x_p < x < x_n$, the charge distribution exhibits a step-like behavior, as illustrated in Fig. 6.7.

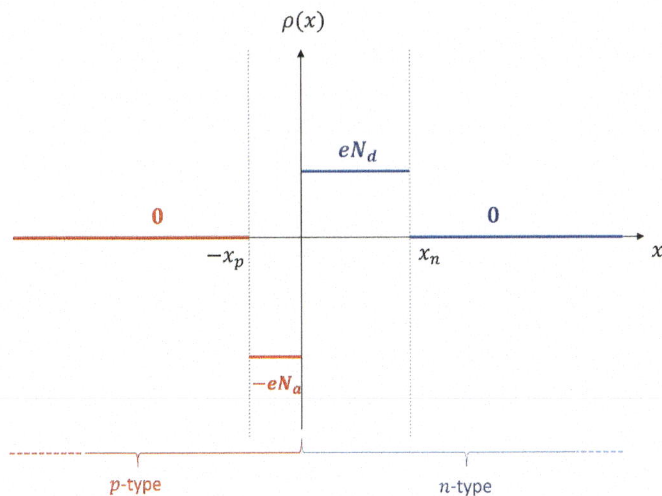

Fig. 6.7 Charge density distribution in a p-n junction estimated with the DA

To summarize, in the DA, the charge density becomes piecewise constant and can be divided into four regions:

$$\rho(x) = \begin{cases} 0, & x < -x_p \\ qN_d, & -x_p < x < 0 \\ -qN_a, & 0 < x < x_n \\ 0, & x > x_n. \end{cases}$$

All the charge density is therefore confined within the space charge region ($-x_p < x < x_n$). Using the **Poisson equation**:

$$\frac{d^2\Phi(x)}{dx^2} = -\frac{\rho(x)}{\varepsilon},$$

it is possible to derive also an expression for the potential $\Phi(x)$ in this region, while in the p-type bulk ($x < -x_p$) and in the n-type bulk ($x > x_n$) the asymptotic conditions are:

$$\Phi(x \to -\infty) = -V,$$

$$\Phi(x \to +\infty) = 0.$$

Let us first consider the n-type region ($x > x_n$) to the right of the junction. For this region, $\Phi(+\infty) = 0$, $n(+\infty) = N_d$, $\rho(+\infty) = 0$, and the electric field is zero:

$$F(+\infty) = -\left.\frac{d\Phi(x)}{dx}\right|_{+\infty} = 0.$$

Hence, Poisson's equation simplifies to:

$$\frac{d^2\Phi(x)}{dx^2} = 0,$$

with:

$$\Phi(+\infty) = 0,$$
$$\Phi'(+\infty) = 0.$$

By integrating two times:

$$\begin{cases} \Phi'(x) = a \\ \Phi(x) = ax + b \end{cases}$$

and by inserting the boundary conditions, the constants result to be:

$$\begin{cases} \Phi'(+\infty) = a = 0 \\ \Phi(+\infty) = b = 0. \end{cases}$$

Thus, the solution for the potential in this region is:

$$\Phi(x > x_n) = 0.$$

Next, consider the region $0 < x < x_n$, where the charge density is $\rho(x) = qN_d$. The continuity conditions for the potential and field at $x = x_n$ are $\Phi(+\infty) = \Phi(x_n)$ and $\Phi'(+\infty) = \Phi'(x_n)$. The Poisson's equation in this region is:

$$\frac{d^2\Phi(x)}{dx^2} = -\frac{qN_d}{\varepsilon},$$

with:

$$\Phi(x_n) = 0,$$
$$\Phi'(x_n) = 0.$$

Integrating twice yields:

$$\begin{cases} \Phi'(x) = -\frac{qN_d}{\varepsilon}x + a \\ \Phi(x) = -\frac{qN_d}{2\varepsilon}x^2 + ax + b. \end{cases}$$

The constants a and b are determined using the boundary conditions:

$$\begin{cases} \Phi'(x_n) = -\frac{qN_d}{\varepsilon}x_n + a = 0 \\ \Phi(x_n) = -\frac{qN_d}{2\varepsilon}{x_n}^2 + ax_n + b = 0. \end{cases}$$

Solving these equations gives:

$$\begin{cases} a = \frac{qN_d}{\varepsilon}x_n \\ b = -\frac{qN_d}{2\varepsilon}{x_n}^2. \end{cases}$$

Substituting these values, the potential in this region is:

$$\Phi(0 < x < x_n) = -\frac{qN_d}{2\varepsilon}(x - x_n)^2,$$

which is parabolic and vanishes at $x = x_n$, as expected.

For the p-type region, in the bulk ($x < -x_p$), Poisson's equation is:

$$\frac{d^2\Phi(x)}{dx^2} = 0$$

with:

$$\Phi(-\infty) = -V,$$
$$\Phi'(-\infty) = 0.$$

Thus, the solution is trivial:

$$\Phi\left(x < -x_p\right) = -V.$$

Within the space charge region ($-x_p < x < 0$), where $\rho(x) = -qN_a$, Poisson's equation becomes:

$$\frac{d^2\Phi(x)}{dx^2} = \frac{qN_a}{\varepsilon},$$

and to guarantee continuity:

$$\Phi\left(-x_p\right) = -V,$$
$$\Phi'\left(-x_p\right) = 0.$$

Integrating twice, the potential in this region results:

$$\Phi\left(-x_p < x < 0\right) = \frac{qN_a}{2\varepsilon}\left(x + x_p\right)^2 - V$$

that for $x = -x_p$ is null.

An example of the Φ(x) profile for a p-n junction in which the p-type material is more heavily doped than the n-type one ($N_a > N_d$) is shown in Fig. 6.8. Within the space-charge region, the potential clearly displays two parabolic branches, one on the p-side and the other on the n-side, that meet smoothly at the junction interface.

It is possible to impose the continuity condition at the junction point. Consider a junction formed between generic p-type and n-type materials. In principle, the two regions (1 and 2) may possess different relative dielectric constants (ε_1 and ε_2). Even in this generalized configuration, the continuity of the displacement vector $\vec{D}$ must hold at the interface ($x = 0$).

Let us consider an infinitesimal cylindrical Gaussian surface crossing the interface perpendicularly, with area dS and height $h \ll \sqrt{dS}$. The flux of a generic vector $\vec{D}$ through this cylinder is:

$$\Phi(\vec{D}) = D_\perp^2 dS - D_\perp^1 dS = 0,$$

This equality holds because no net volume charge can be enclosed within the infinitesimal volume. Let us suppose that the junction is imperfect (real junction) and that there is a free charge at the interface, corresponding not to a polarization charge (which is irrelevant in Gauss's law for $\vec{D}$) but to a mobile free charge $q = \sigma dS$

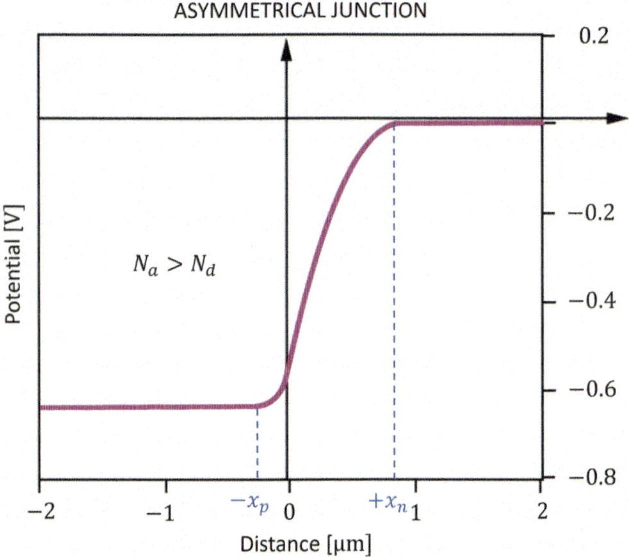

Fig. 6.8 Example of the potential profile, expressed in V, for a p-n junction where $N_a > N_d$ as a function of the distance x from the center of the junction ($x = 0$), expressed in μm

trapped in surface states. Even in this situation, the product of a finite free charge density ($\rho \neq 0$) and the vanishing infinitesimal volume tends to zero, giving a negligible contribution. More rigorously, the normal component of the displacement field satisfies the boundary condition:

$$\vec{D}^n_\perp - \vec{D}^p_\perp = \sigma_{\text{free}},$$

where σ_{free} is the free surface charge density at the interface. In the ideal case $\sigma_{\text{free}} = 0$, the displacement field is continuous, whereas in real metal–semiconductor junctions a finite surface charge leads to a discontinuity of $\vec{D}_\perp$.

In the ideal case, where no free interfacial charge is present, the condition $\Phi(\vec{D}) = 0$ directly implies the continuity of $\vec{D}$ at $x = 0$. Since $\vec{D} = \varepsilon\vec{F}$ (where $\vec{F}$ is the built-in electric field), the continuity condition becomes:

$$\varepsilon_1 F_1 = \varepsilon_2 F_2.$$

For the electric field $\vec{F}$, we have on the p-side:

$$F_1 = -\frac{\partial \Phi}{\partial x} = -\frac{\partial}{\partial x}\left[\frac{qN_a}{2\varepsilon_1}\left(x + x_p\right)^2 - V\right] = -\frac{qN_a}{\varepsilon_1}\left(x + x_p\right),$$

and on the n-side:

$$F_2 = -\frac{\partial \Phi}{\partial x} = -\frac{\partial}{\partial x}\left[-\frac{qN_d}{2\varepsilon_2}(x - x_n)^2\right] = \frac{qN_d}{\varepsilon_2}(x - x_n).$$

Evaluating these expressions at $x = 0$ and substituting them into the continuity condition yields:

$$-\varepsilon_1 \frac{qN_a}{\varepsilon_1}\left(x_p\right) = -\varepsilon_2 \frac{qN_d}{\varepsilon_2}(x_n),$$

which simplifies to:

$$N_a x_p = N_d x_n.$$

This is the **neutrality condition**. By multiplying both sides by A (the cross-sectional area at the interface between the p-type and n-type materials) where Ax_n and Ax_p represent the volumes on the n- and p-sides, respectively (see Fig. 6.9), the equation expresses the balance of charge in the two regions. Notably, the dependence on the dielectric constants ε_1 and ε_2 vanishes in this final expression.

Thus, the charge stored in the n-type region compensates the charge stored in the p-type region. Let us now derive explicit expressions for x_n and x_p, assuming that both semiconductors have the same dielectric constant ε. From the charge neutrality condition:

$$x_n = \frac{N_a}{N_d} x_p.$$

At equilibrium, the electrostatic potential must be continuous across the space-charge region, so we write:

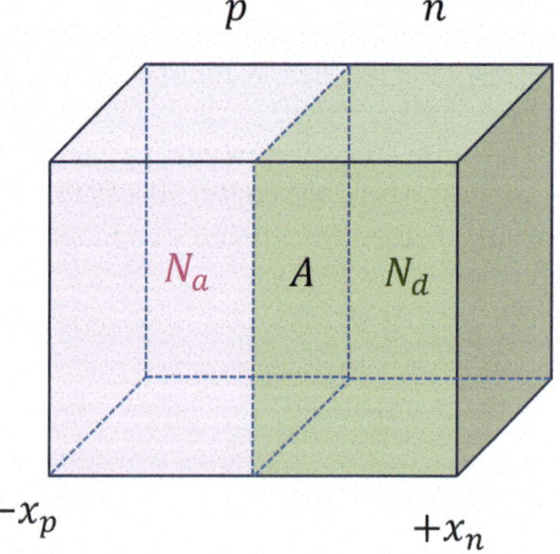

Fig. 6.9 $N_a A x_p$ is the charge in the parallelepiped of volume Ax_p, while $N_d A x_n$ is the charge in the parallelepiped of volume Ax_n

$$\frac{qN_a}{2\varepsilon}{x_p}^2 - V = -\frac{qN_d}{2\varepsilon_2}{x_n}^2$$

and:

$$V = \frac{q}{2\varepsilon}N_a{x_p}^2\left(1 + \frac{N_d{x_n}^2}{N_a{x_p}^2}\right)$$

Substituting x_n in terms of x_p gives:

$$V = \frac{q}{2\varepsilon}N_a{x_p}^2\left(1 + \frac{N_d\left(\frac{N_a}{N_d}x_p\right)^2}{N_a{x_p}^2}\right) = \frac{q}{2\varepsilon}N_a{x_p}^2\left(1 + \frac{N_a}{N_d}\right)$$

From this relation, the depletion widths are obtained as:

$$x_p = \sqrt{\frac{2\varepsilon V}{qN_a\left(1 + \frac{N_a}{N_d}\right)}}, \quad x_n = \sqrt{\frac{2\varepsilon V}{qN_d\left(1 + \frac{N_d}{N_a}\right)}}.$$

In the limit $N_a \gg N_d$:

$$x_p \to 0, \quad x_n \approx \sqrt{\frac{2\varepsilon V}{qN_d}}.$$

Thus, the region on the p-side collapses into an effectively infinitesimal surface of negative charge. Therefore, in an asymmetric junction (Fig. 6.10), the more heavily doped region always corresponds to the side with the shorter depletion width. This behavior follows directly from charge neutrality condition: only a small volume is required to compensate the charge on the lightly doped side.

This asymmetric case is particularly important, as it reflects the limiting behavior of a Schottky contact (a metal-semiconductor junction) discussed in Sect. 6.5.

6.4 The Debye Length Approximation

The DA cannot always be applied, particularly in situations where the full expression for the charge density must be used. In such cases, the potential profile obtained from Poisson's equation becomes significantly more complex than the parabolic form predicted by the DA. Let us consider the previously derived expressions for the charge density in p- and n-type materials:

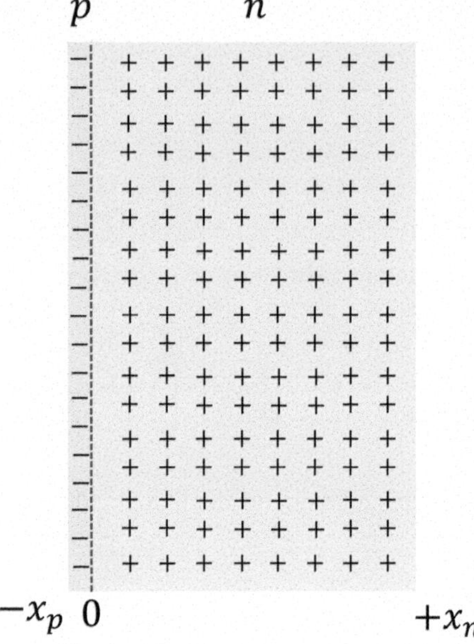

Fig. 6.10 Asymmetric junction with high doping in the p-type region ($N_a \gg N_d$), where $x_p \to 0$ and the p-type region collapses into a surface

$$\rho(x) = \begin{cases} -qN_a + qN_a e^{\frac{-q(V_{BI}+\Phi(x))}{K_BT}}, & p-type \\ qN_d - qN_d e^{\frac{q\Phi(x)}{K_BT}}, & n-type. \end{cases}$$

Now consider the n-type region and insert $\rho(x)$ into Poisson's equation:

$$\frac{d^2\Phi(x)}{dx^2} = -\frac{\rho(x)}{\varepsilon} = -\frac{qN_d}{\varepsilon} + \frac{qN_d e^{\frac{q\Phi(x)}{K_BT}}}{\varepsilon}.$$

This is a strongly nonlinear differential equation that, in some regions, admits no closed-form analytical solution and must be solved numerically. Nevertheless, it is possible to identify a regime in which the equation can be linearized, leading to a full analytical solution that connects smoothly to the parabolic behavior observed within the depletion region.

Consider a region to the right of the junction where the potential is sufficiently small, satisfying the condition:

$$|\Phi(x)| \ll \frac{k_BT}{q}$$

(where q=+e and which is distinct from the DA condition, for which V$\gg$ k_BT/q). In this limit, the exponential term can be expanded to first order, yielding:

$$\frac{d^2\Phi(x)}{dx^2} \approx -\frac{qN_d}{\varepsilon}\left(1 - 1 - \frac{q\Phi(x)}{K_BT}\right) = \frac{q^2N_d}{\varepsilon K_BT}\Phi(x).$$

Defining the constant:

$$\frac{\varepsilon k_B T}{q^2 N_d} = {L_D}^2,$$

the **Debye length L_D** is introduced as:

$$L_D = \sqrt{\frac{\varepsilon K_B T}{q^2 N_d}}.$$

So, the Poisson's equation simplifies to:

$$\frac{d^2\Phi(x)}{dx^2} = \frac{\Phi(x)}{{L_D}^2}$$

whose general solution is:

$$\Phi(x) = Ae^{-\frac{x}{L_D}} + BAe^{\frac{x}{L_D}}.$$

For the potential to remain finite as $x \to \infty$, the term BAe^{x/L_D} must vanish, leaving:

$$\Phi(x) = Ae^{-\frac{x}{L_D}}$$

and the Debye length provides a measure of the characteristic length over which the potential decays. Applying an analogous analysis to the p type region (left of the junction), the solution takes the form:

$$\Phi(x) = Ce^{\frac{x}{L_A}} - V,$$

where the Debye length for the p-type material, is defined as:

$$L_A = \sqrt{\frac{\varepsilon K_B T}{q^2 N_a}}.$$

As illustrated in Fig. 6.11, the potential $\Phi(x)$ is divided into six distinct regions. Near the junction (the central regions), the potential retains a parabolic shape, consistent with the DA. In the intermediate "**screening regions**" (shown in yellow), the influence of the potential is non-negligible, and no analytical solution exists; here, the potential decays exponentially over a distance of the order of the Debye length. Beyond these regions, the potential decays asymptotically as e^{-x/L_D} on the n-side and as $e^{x/L_A} - V$ on the p-side.

Let us now compare the Debye length with the depletion widths x_n and x_p. For instance, on the n-side we previously found:

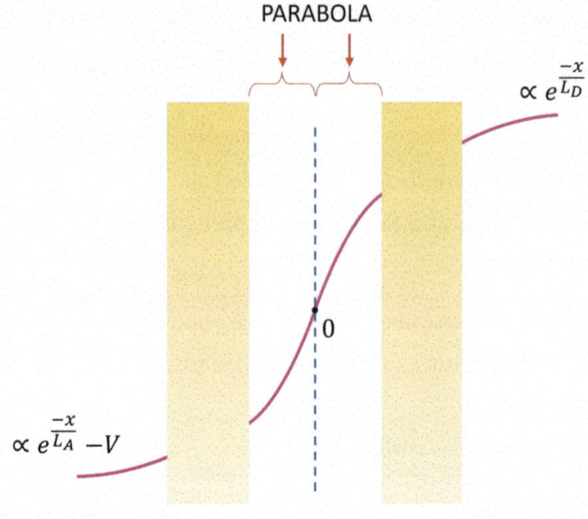

Fig. 6.11 Potential profile Φ(x) as a function of the distance from the junction. In the central regions the shape is parabolic, in the yellow regions it is governed by the Debye lengths, in the external regions it is proportional to e^{-x/L_D} on the n-side and to $e^{x/L_A} - V$ on the p-side

$$x_n \approx \sqrt{\frac{2\varepsilon V}{qN_d}}.$$

Thus:

$$\frac{L_D}{x_n} = \frac{\sqrt{\frac{\varepsilon K_B T}{qN_d}}}{\sqrt{\frac{2\varepsilon V}{q^2 N_d}}} = \sqrt{\frac{K_B T}{2qV}}.$$

If $V \gg K_B T/q$, then:

$$\frac{L_D}{x_n} \to 0, \; \frac{L_A}{x_p} \to 0,$$

which corresponds to the DA, the limiting case of a more general theory that provides the complete expression for the potential. When the potential is not much larger than the thermal voltage, the points $-x_p$ and x_n defined in the DA behave as if "magnified" under a lens, revealing a finite spatial structure governed by the Debye length rather than collapsing to idealized boundaries. In this regime, the Debye length becomes physically meaningful.

It is essential to emphasize that, as will be discussed later, the key parameter determining the nanostructure is not the Debye length (L_D or L_A), but the depletion width (x_p and x_n). The Debye length becomes relevant only when the potential is small and comparable to the thermal potential (V~$K_B T/q$); for large potentials, the grain radius should not be compared with the Debye length.

In summary, the Debye model represents a refinement of the DA providing additional detail in regions where the DA is insufficient.

6.5 Metal-Semiconductor Contact

Let us now examine the case in which the n-type region of a p-n junction is so heavily doped ($N_d \gg N_a$) that the p-type side effectively collapses into a thin sheet of charge. To analyze the system through the DA, we must impose the continuity of the potential and of the electric displacement field at the interface, which leads to the neutrality condition:

$$x_p N_a = x_n N_d .$$

Interestingly, this relation remains valid even in situations where the continuity of the displacement field no longer holds due to the appearance of surface charges rather than polarization charges.

Let us consider the case of an n-type semiconductor in contact with a metal (whose Fermi level coincides with its Fermi energy) as shown in Fig. 6.12.

When the two materials are isolated, a vacuum level E_{VAC} can be defined as the potential energy of an electron that is completely free from the material, having zero kinetic energy and positioned just outside the surface. This level serves as the common reference for comparing their electronic properties and is critical for determining energy barriers at material interfaces.

For a semiconductor, the quantities of interest are:

- the **electron affinity** X_S, the energy difference between the conduction band minimum and the vacuum level, or the energy required to move an electron from the conduction band to the vacuum level;
- the **extraction potential** I_S, the energy difference between the valence band maximum and the vacuum level, or the energy required to move an electron

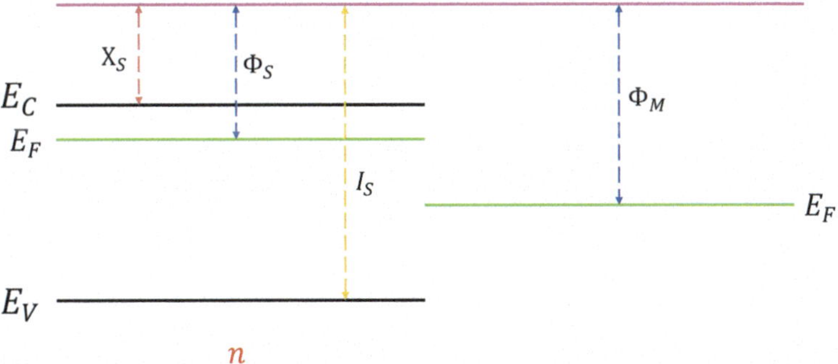

Fig. 6.12 Energy band structure of an n-type semiconductor and a metal, where $\Phi_S < \Phi_M$, before contact. The energy of the conduction band minimum is E_C, the energy of the valence band maximum is E_V, the Fermi levels are E_F, and the vacuum level is represented by the pink line. Φ_S and Φ_M are the work functions of the semiconductor and the metal, respectively; X_S is the electron affinity of the semiconductor; I_S is the extraction potential of the semiconductor

from the valence band to the vacuum level (it is the quantity involved in the photoelectric effect)

- the **work function Φ_S**, the minimum energy required to remove an electron from its surface, specifically from the Fermi level to the vacuum level (it is a statistical concept).

For a metal, these distinctions collapse: the **work function Φ_M** coincides with both the electron affinity and the extraction potential.

In the configuration shown in Fig. 6.12, the Fermi level of the semiconductor lies closer to the vacuum level than that of the metal, implying that the semiconductor has a smaller work function Φ_S compared to the metal's work function Φ_M.

When two materials are brought into contact with $\Phi_S < \Phi_M$, the **Schottky model** applies. At equilibrium, the difference in Fermi levels (ΔE_F) vanishes, meaning that the Fermi levels of both materials must coincide. Upon contact, electrons flow from the semiconductor toward the metal, re-establishing this alignment. The electron flux from the semiconductor to the metal is proportional to an exponential factor of the form $e^{-q\Phi_S/k_BT}$ (according to Richardson's law), which exceeds the electron flux in the opposite direction (from the metal to the semiconductor) characterized by the exponential dependence $e^{-\Phi_M/k_BT}$. Consequently, the net electron transfer occurs from the semiconductor to the metal, causing the semiconductor to acquire a positive potential relative to the metal. This reduces the potential energy of the electrons (because the electrostatic potential enters with a negative sign), shifting all the semiconductor bands downward until the Fermi levels align (see Fig. 6.13).

Thus, the entire near-surface region of the semiconductor moves farther away from the Fermi level, leading to a reduction in the local electron concentration. In fact, band bending typically signals the formation of a depletion region, where electrons have migrated to the opposite side. Far from the junction, in the bulk, the band structure

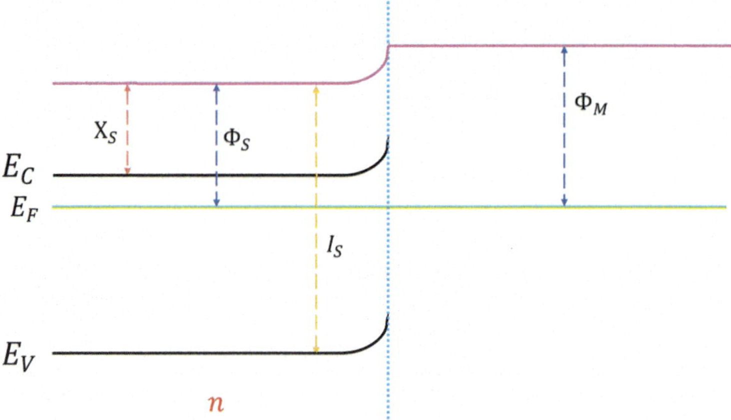

Fig. 6.13 Energy band structure of a n-type semiconductor and a metal, where $\Phi_S < \Phi_M$, after contact. On the surface appears a band bending

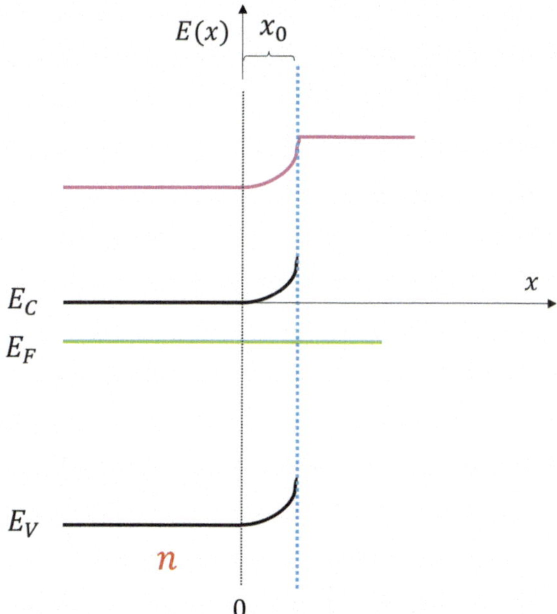

Fig. 6.14 Magnification of the band bending at the interface between a n-type semiconductor and a metal, where x_0 indicates the width of the band bending region

remains unchanged. Let the width of the band-bending region be denoted by x_0, as illustrated in the magnification in Fig. 6.14.

The zero of energy is defined in the bulk of the n-type semiconductor, corresponding to $E_c(x \to -\infty)$. Hence, the left-hand side of the diagram represents the fully n-type bulk region. All band bending occurs within the depletion region, where the electric field depletes the space-charge layer. The work function Φ_S remains the same as in the isolated semiconductor (see Fig. 6.13). As shown in Fig. 6.15, the electron affinity X_S remains constant and follows the band bending, while the energy difference between the conduction band minimum in the bulk and the Fermi level can be written as:

$$E_c(\text{bulk}) - E_F = \xi$$

With these parameters, the Schottky barrier associated with the contact can be determined precisely. From Fig. 6.15, the difference between the conduction band edge at the surface and in the bulk is equal to the electrostatic potential energy:

$$E_c(\text{surface}) - E_c(\text{bulk}) = qV$$

which is the total band bending, as usually defined. Moreover, Fig. 6.15 shows that the difference between the metal work function and the semiconductor electron affinity corresponds exactly to the Schottky barrier height:

$$\Phi_M - X_S = S_B.$$

Additionally, the Schottky barrier height can also be expressed as:

$$S_B = qV + \xi.$$

Thus, we obtain:

$$qV + \xi = \Phi_M - X_S,$$

or equivalently:

$$qV = \Phi_M - (X_S + \xi).$$

Since $X_S + \xi = \Phi_S$, the result reduces to:

$$qV = \Phi_M - \Phi_S.$$

This shows that, when a metal is brought into contact with a semiconductor, a surface barrier (the Schottky barrier) forms, given by the difference between the work function of the metal and the electron affinity of the semiconductor. The total band bending is determined by the difference between the work function of the metal and that of the semiconductor.

Let us examine more closely the behavior of the electric field at the interface, as illustrated in Fig. 6.16. The tangential component of the electric field $\vec{F}$ must be continuous across the interface. By evaluating the circulation of the field along a rectangular contour having two sides parallel to the interface (each of length *dl*) and two perpendicular sides of infinitesimal height, one obtains:

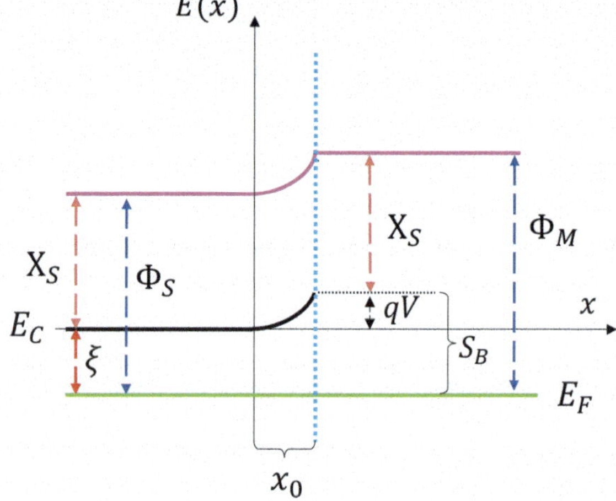

Fig. 6.15 Magnification of the band bending at the interface between a *n*-type semiconductor and a metal, where Φ_S and X_S are indicated

$$\oint \vec{F} \cdot d\vec{l} = 0.$$

This confirms that the electrostatic field is conservative. Expanding the circulation gives:

$$F^1_{//}dl - F^2_{//}dl = 0,$$

hence:

$$F^1_{//} = F^2_{//}.$$

Thus, the tangential component of the electric field is continuous. This holds in general because no time-varying magnetic field is present at the interface, in accordance with Faraday's law. At equilibrium, however, the electric field inside the metal is zero, implying that its tangential component must vanish at the boundary. Consequently, the electric field in the semiconductor must be strictly perpendicular to the junction, producing a first-order discontinuity in the field at the interface (Fig. 6.17).

When the field reaches the metal, it is immediately neutralized. The normal component $D_{\perp}$ of the electric displacement field $\vec{D}$ is instead discontinuous, as a real surface charge density σ exists at the interface (Fig. 6.18). This discontinuity is not due to polarization charge but to a genuine accumulation of electrical charge (since the polarization charge does not affect the flow of $\vec{D}$). Thus, a non-polarization charge must always be present at the interface between the metal and the semiconductor because the electric field must be discontinuous according to basic physical principles. This is referred to as the so-called double layer.

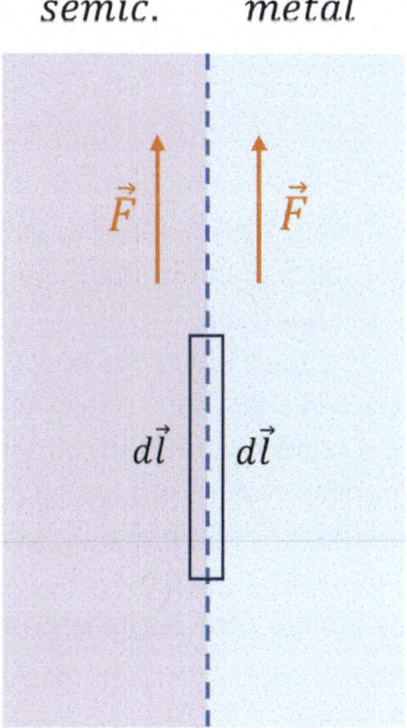

Fig. 6.16 Interface of a semiconductor and a metal

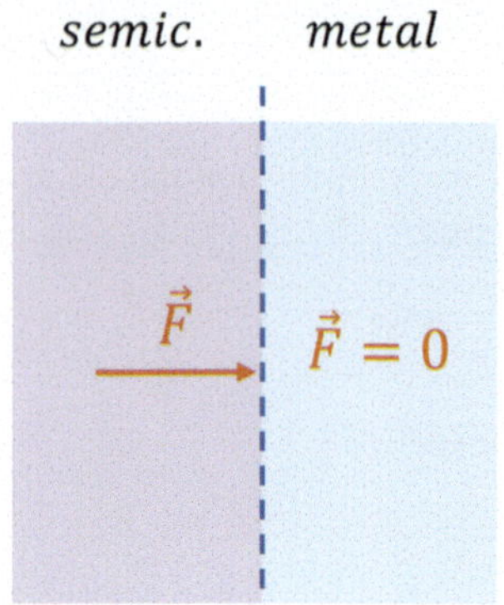

Fig. 6.17 First order discontinuity of the field at the interface of a semiconductor and a metal

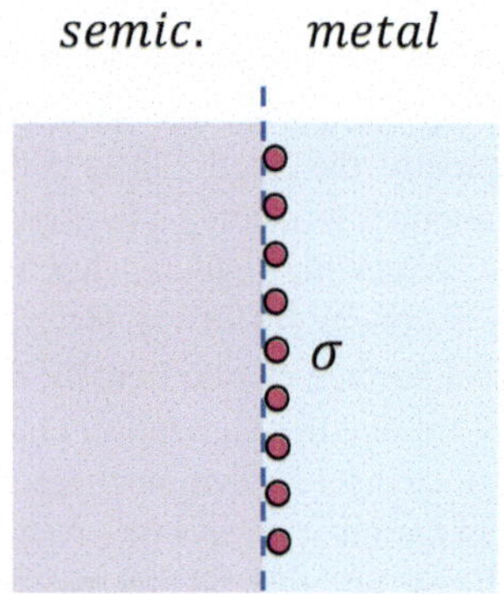

Fig. 6.18 Surface charge distribution σ at the interface between a semiconductor and a metal

Being N_S the density of surface states (typically measured in m^{-2}), the surface charge density is given by:

$$\sigma_S = qN_S,$$

where q is the single electron charge. Since these surface states are generally electron-trapping, the resulting σ_S is negative. This surface charge contributes significantly (far more than originally anticipated by Schottky) to determining the barrier height at the interface.

Fig. 6.19 illustrates how this negative surface charge is compensated by a positive space charge in the semiconductor over the region $0 < x < x_0$, caused by the conduction band bending. Electrons escaping into the metal leave behind ionized donors (of density N_d), producing a positive volume charge. The region $x < 0$ (in the semiconductor bulk) remains overall neutral. At the relevant temperatures, essentially all donors are ionized ($N_d \approx N_d^+$).

Charge neutrality therefore requires:

$$N_d x_0 = N_S,$$

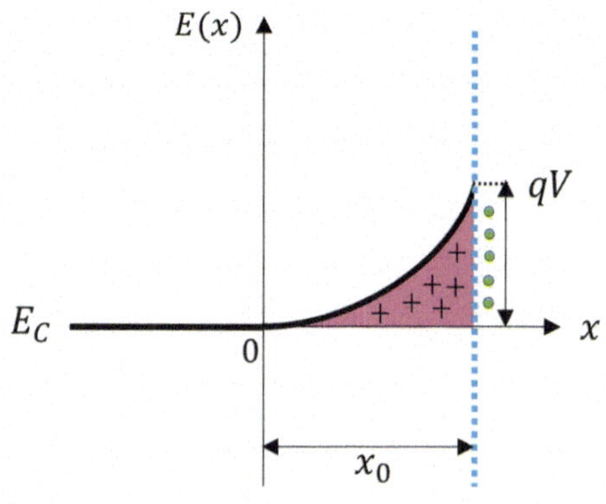

Fig. 6.19 Neutrality condition between the negative surface charge σ at a semiconductor/metal interface and a positive volume charge in the semiconductor region $0 < x < x_0$; at $x < 0$ (in the semiconductor bulk) the region is neutral

which is the **neutrality condition** equating the positive volume charge of donors to the negative surface charge of acceptors (electrons trapped in the surface states). Note the dimensional consistency: $[N_d] = \mathrm{m}^{-3}$ and $[N_S] = \mathrm{m}^{-2}$.

In the depletion region, Poisson's equation reads:

$$\frac{d^2\Phi(x)}{dx^2} = -\frac{qN_d}{\varepsilon},$$

with the asymptotic conditions:

$$\Phi(x \to -\infty) = 0, \quad \Phi(0) = 0,$$

and consequently:

$$\Phi'(0) = 0.$$

Integrating the equation:

$$\frac{d\Phi(x)}{dx} = -\frac{qN_d}{\varepsilon}x + A,$$

$$\Phi(x) = -\frac{qN_d}{2\varepsilon}x^2 + Ax + B.$$

The boundary conditions give $A = 0$ and $B = 0$ and the final expression for potential results in a parabolic shape:

$$\Phi(x) = -\frac{qN_d}{2\varepsilon}x^2.$$

At the semiconductor surface ($x = x_0$):

$$\Phi(x_0) = -\frac{qN_d}{2\varepsilon}{x_0}^2 = -V_S.$$

Therefore, the depletion width is:

$$x_0 = \sqrt{\frac{2\varepsilon V_S}{qN_d}}.$$

which is formally identical to the depletion width of a p-n junction where one side (here replaced by a metal) behaves as a heavily doped, quasi-metallic region. When one side of a p-n junction is sufficiently degenerate, its depletion region collapses to an effectively two-dimensional layer, consistent with metallic behavior.

Consider now an n-type semiconductor in contact with a metal. If the metal work function is smaller than that of the semiconductor ($\Phi_M < \Phi_S$), as in Fig. 6.20, electrons flow from the metal into the semiconductor upon contact.

When the materials are brought into contact, the situation is reversed compared to the previous case: the electron flow from the metal to the semiconductor becomes dominant. As a result, the Fermi level of the metal will shift downward, and the energy levels will bend in the opposite direction compared to the Schottky contact, as shown in Fig. 6.21, forming an **accumulation layer** (instead of a space charge region). As a result, the probability of finding electrons in this region increases. The newly created region thus acts as a source of electrons.

This means that, by applying a positive potential on the left and a negative potential on the right-hand side of the junction, the current flows normally, and electrons encounter no barrier (flowing from right to left). However, by applying a negative potential on the left and a positive potential on the right-hand side of the junction, the current still flows, because electrons in the accumulation layer are driven by the electric field towards the positive region. This indicates that this type of contact is

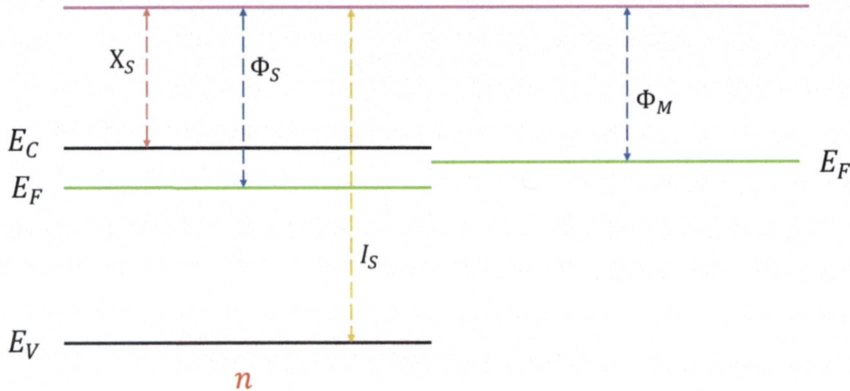

Fig. 6.20 n-type semiconductor and metal (with $\Phi_M < \Phi_S$) before contact

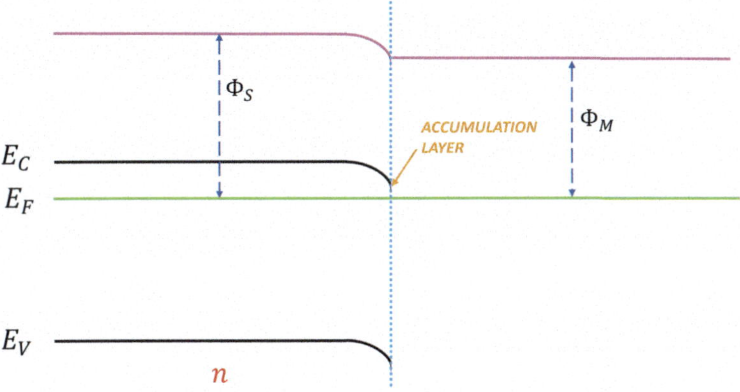

Fig. 6.21 Contact between a *n*-type semiconductor and a metal with. $\Phi_M < \Phi_S$. At the interface there is now an accumulation layer instead of a space charge region

"symmetrical" for current flow and is referred to as an **Ohmic contact**. In contrast, the Schottky contact ($\Phi_M > \Phi_S$ for a n-type material) acts as a rectifier.

For p-type semiconductors, the situation is reversed. Let us consider the case of a p-type semiconductor in contact with a metal. First, we examine the situation where $\Phi_S < \Phi_M$, as shown in Fig. 6.22.

When the two materials are brought into contact, electrons flow from the metal into the semiconductor. For a p-type semiconductor, the analysis is most naturally carried out in terms of holes. The electron transfer from the metal reduces the hole density near the interface, producing a separation between the valence band and the Fermi level. This leads to the formation of a **hole depletion region** (Fig. 6.23).

The resulting band bending constitutes a **Schottky barrier for holes**, the majority carriers. Under forward bias (positive potential applied to the semiconductor side and negative to the metal) holes can surmount the barrier and current flows. Under reverse

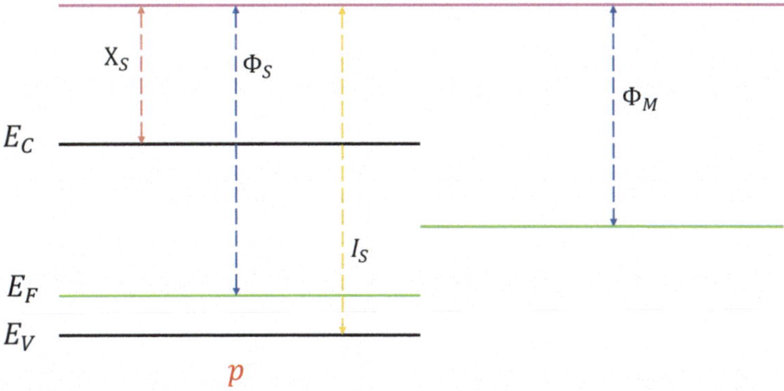

Fig. 6.22 p-type semiconductor and metal (with $\Phi_M < \Phi_S$) before contact

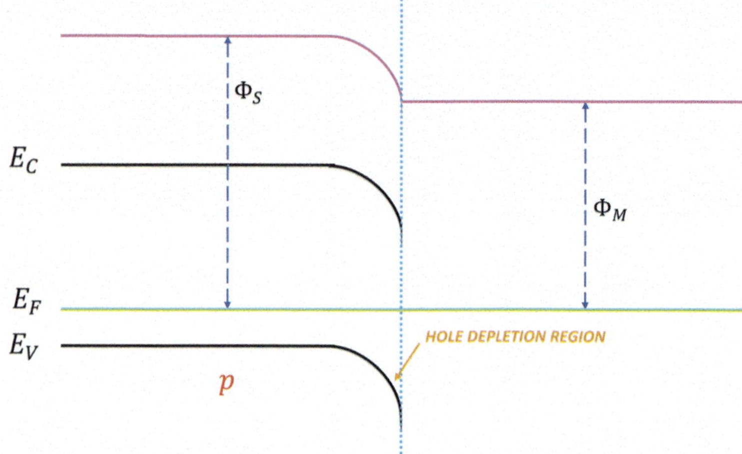

Fig. 6.23 Contact between a p-type semiconductor and a metal with $\Phi_M < \Phi_S$. A Schottky barrier for holes is formed at the interface

bias, however, the applied potential increases the barrier height for holes (i.e., the potential well for electrons), suppressing hole transport and effectively blocking the current.

In contrast, when the metal work function exceeds that of the p-type semiconductor ($\Phi_M > \Phi_S$), as in Fig. 6.24, the band alignment at equilibrium results in an upward shift of the valence band near the interface (positive band bending). This produces a **hole accumulation layer** (Fig. 6.25), rather than a depletion region.

In this case, an Ohmic contact is formed, with an accumulation region for holes at the interface, allowing current to flow in both directions, regardless of the potential polarity on either side of the junction. As will be discussed in the next chapters, our sensors typically involve n-type semiconductors.

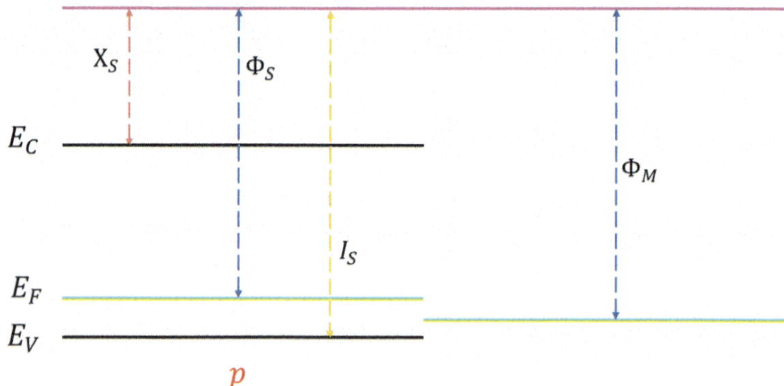

Fig. 6.24 p-type semiconductor and metal (with $\Phi_M > \Phi_S$) before contact

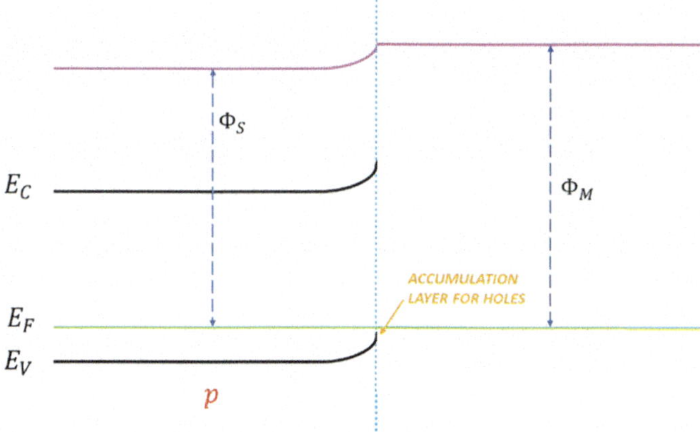

Fig. 6.25 Contact between a p-type semiconductor and a metal with $\Phi_M > \Phi_S$. A hole accumulation layer is formed at the interface

Table 6.1 Relation between semiconductor types, work-functions, and contact behavior

Semiconductor type	Work-function relation	Contact type
n	$\Phi_M > \Phi_S$	Schottky
n	$\Phi_M < \Phi_S$	Ohmic
p	$\Phi_M > \Phi_S$	Ohmic
p	$\Phi_M < \Phi_S$	Schottky

These four configuration are summarized in Table 6.1:

The reversal between n-type and p-type cases is evident. A practical way to determine the nature of the contact is as follows:

1. draw the band diagrams of the isolated metal and semiconductor;
2. upon contact, align the Fermi levels by shifting the metal's Fermi level upward or downward as required;
3. finally, bend the semiconductor bands while preserving their internal energy separations.

The resulting equilibrium band diagram directly reveals whether the interface behaves as a Schottky or Ohmic contact.

6.6 Bardeen Model

The contact between a semiconductor and a metal can be measured using the **Kelvin Probe method**, which employs an oscillating probe to determine the contact potential difference via capacitive measurements, i.e., the Volta effect. This potential difference directly corresponds to the difference between the two Fermi levels and thus represents the work function difference $\Delta\Phi$ between the two materials, referenced to a common vacuum level. Although the Kelvin probe is a complex and delicate system to implement, it provides a reliable means to measure $\Delta\Phi$ independently of the metal in contact with the semiconductor.

However, this method exhibits an experimental limitation. Specifically, $\Delta\Phi = \Phi_M - \Phi_S$ (in eV) has been experimentally measured for contacts between n-type doped silicon (Si^n) and metals such as gold (Au), platinum (Pt), and silver (Ag). Observations reveal that the Schottky barrier ($S_B = qV + \xi$) remains nearly constant despite changes in the metal, indicating that it is independent of variations in the metal's work function. In other words:

$$\frac{dS_B}{d\Phi_M} = 0.$$

This behavior is particularly pronounced in **covalent semiconductors**, where the Schottky barrier does not follow the expected model, and

$$S_B \neq \Phi_M - X_S.$$

This phenomenon was first noted by Bardeen in 1947, who proposed that the presence of surface states, which influence the effective barrier height, could explain this observation. This effect is known as **Fermi-level pinning** and is described by the theory of **Metal-Induced Gap States (MIGS)**. MIGS are interface states at the semiconductor-metal junction that regulate the Schottky barrier height.

The **Bardeen model** provides a theoretical framework for this phenomenon, focusing on the interface. Consider an n-type semiconductor in contact with a metal or a gas (i.e., a material with a higher work function). The semiconductor's energy band structure can be represented as shown in Fig. 6.26. In this model, a distribution of forbidden states appears within the band gap, as the surface acts as a defect, breaking the crystal symmetry and the Bloch periodicity, thus generating surface states. These surface states are classified as either donor-type or acceptor-type. While the exact number of surface states cannot be determined a priori, an equilibrium neutrality condition can be established at the surface.

A **neutral level**, denoted as $\boldsymbol{\Phi_0}$, is defined such that states above Φ_0 are acceptor-like and states below Φ_0 are donor-like. Defining Φ_0 is crucial because, regardless of the unknown doping at the surface or the precise nature of the surface states, when the Fermi level coincides with the neutral level ($E_F \equiv \Phi_0$), all states below E_F are filled, and all states above are empty. This condition holds exactly at absolute zero and is a good approximation at room temperature. Consequently, all acceptor states at the surface are empty, and all donor states are filled, resulting in a neutral surface.

The position of Φ_0 relative to E_F determines whether the surface carries a positive or negative charge. To understand Schottky barrier formation according to Bardeen's model, consider an n-type semiconductor with intrinsic Φ_S and Φ_0 in contact with a metal of work function $\Phi_M > \Phi_S$, forming a Schottky contact. This results in band

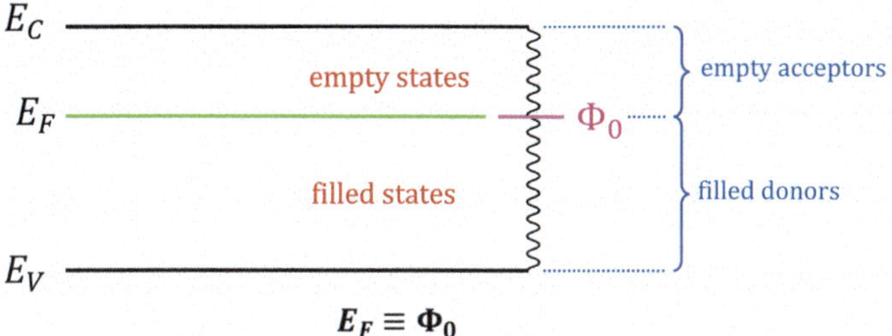

Fig. 6.26 Case in which $E_F \equiv \Phi_0$ in a contact between an *n*-type semiconductor and a metal (or a gas)

bending, as illustrated in Fig. 6.27. All energy levels shift upward by the same amount due to the band bending, including Φ_0. However, Φ_0 cannot coincide with the Fermi level because the surface must not be neutral to maintain a finite surface charge, $N_S = N_d x_0$, required to counterbalance the positive charge of the depletion region. Therefore, Φ_0 lies slightly below E_F, producing a **negative surface charge** in the indicated region of Fig. 6.27b, resulting from filled acceptor states (occupied because they lie below E_F).

By multiplying the positive energy difference $(E_F - \Phi_0)$ with the density of surface states per unit energy, N_{SS} (measured in $eV^{-1} \cdot m^{-2}$), one obtains the surface charge:

$$N_S = N_{SS}(E_F - \Phi_0),$$

which governs the Schottky barrier. Equating this to the total surface charge:

$$N_{SS}(E_F - \Phi_0) = N_S = x_0 N_d$$

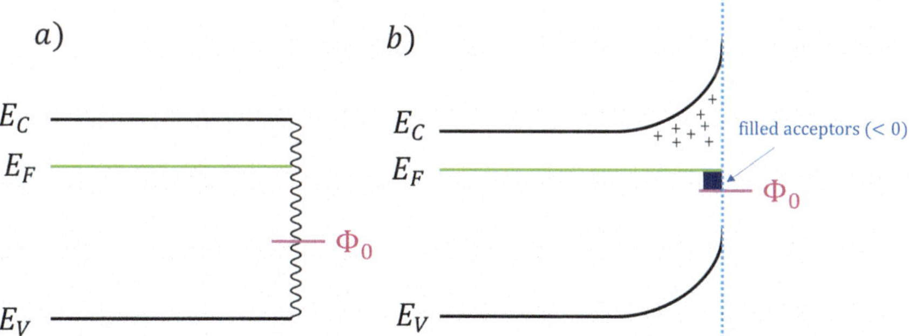

Fig. 6.27 **a** n-type semiconductor; **b** band bending after the contact with a metal (or a gas) with $\Phi_M > \Phi_S$

where x_0 and N_d are finite, implying that N_S must also be finite. Thus, if $N_{SS} \to \infty$ (i.e., it is very large), then for N_S to remain finite, $\mathrm{E_F} \equiv \Phi_0$. The Schottky barrier is defined as:

$$S_B = E_C(\text{surface}) - E_F,$$

where E_C(surface) is the conduction band edge at the surface. Defining:

$$L_0 = E_C - \Phi_0,$$

as shown in Fig. 6.28, we see that L_0 is constant, depending solely on material properties (since Φ_0 moves with the energy bands). Therefore, when $\mathrm{E_F} \equiv \Phi_0$:

$$S_B = E_C(\text{surface}) - \Phi_0 = L_0.$$

This means that the Schottky barrier is **pinned** and is solely determined by the position of the neutral level Φ_0. In this scenario, the surface states dictate the barrier height. The barrier height cannot be changed because the Fermi level is forced to remain infinitesimally close to the neutrality level, due to the very high density of states per unit energy. Changing the material (and thus Φ_M) does not affect the Schottky barrier because it is pinned.

This phenomenon has been experimentally verified, for example, in materials with high N_{SS}, such as Si and Ge, (**covalent semiconductors**). But what happens when N_{SS} is lower? Using the same expression:

$$N_S = N_{SS}(E_F - \Phi_0),$$

since N_S must remain finite, $E_F - \Phi_0$ can also be finite, leading to:

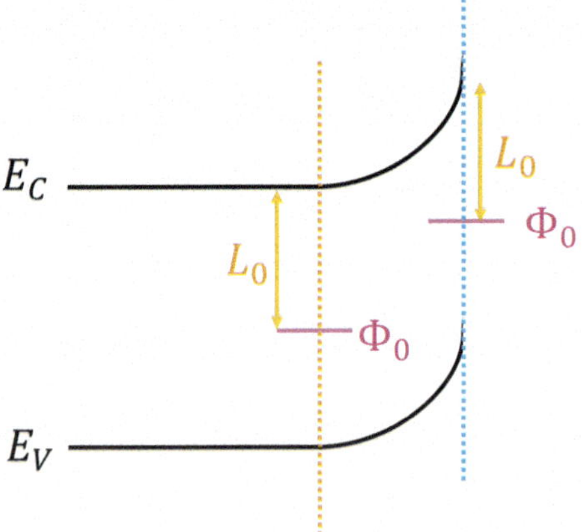

Fig. 6.28 Graphical representation of L_0

$$E_F \neq \Phi_0.$$

This allows the band bending to vary, with E_F moving away from the pinned level. The barrier height can now be modulated by changing the work function Φ_M of the contacting material. This **Fermi-level unpinning** is crucial for the gas-sensing mechanism. In **ionic semiconductors** (e.g., SnO_2 or TiO_2), N_{SS} is low, enabling detection of environmental gas variations; these materials are used in the fabrication of gas sensors.

The following argument provides a physical interpretation of the pinning phenomenon rather than a strict microscopic derivation.

In covalent semiconductors, the uncertainty principle imposes $\Delta x \Delta p \sim \hbar/2$. Owing to the crystalline structure, Δp is directly connected to the kinetic-energy contribution $(\Delta p^2/2m)$, so the uncertainty relation can equivalently be expressed in terms of energy, involving the product $\Delta x \Delta E$. An increase in ΔE correspondingly reduces Δx, and vice versa. In covalent bonding, electrons are shared within an extended electronic cloud throughout the molecular framework (as in Si and Ge), resulting in a highly delocalized spatial distribution Δx and, consequently, a relatively small ΔE. Hence:

$$N_{SS} \sim \frac{N_S}{\Delta E} \to \infty,$$

leading to the pinning phenomenon.

For ionic semiconductors (e.g., SnO_2, TiO_2, WO_3), electrons are more spatially localized due to the ionic bond, resulting in smaller localization in energy (i.e., higher ΔE), low N_{SS}, and $E_F > \Phi_0$, leading to no pinning. Changes in the environment therefore alter the Schottky barrier height, which is the operational principle for gas sensors. The distinction between the two cases is summarized in Table 6.2.

Observing Fig. 6.29, the barrier height and related quantities can be summarized. Previously derived expressions for S_B include:

$$S_B = E_C(\text{surface}) - E_F,$$

$$S_B = \Phi_\mathrm{M} - \mathrm{X}_S,$$

$$S_B = \Phi_\mathrm{M} - \mathrm{X}_S.$$

Additionally, from previous calculations:

Table 6.2 Pinning and unpinning of the Fermi level depending on the bond type and material

Bond type	Fermi level	Materials (Example)
Ionic	Unpinned	SnO_2, TiO_2, WO_3
Covalent	Pinned	Si, Ge

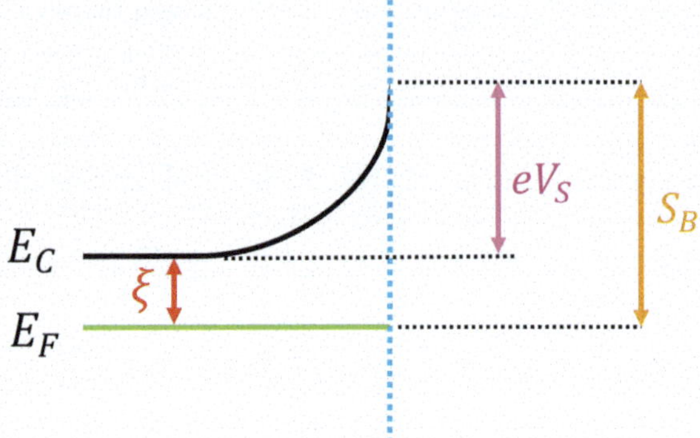

Fig. 6.29 Schottky barrier height and related quantities

$$N_d x_0 = N_S, \quad V_S = \frac{eN_d}{2\varepsilon} {x_0}^2.$$

Combining the two:

$$V_S = \frac{eN_d}{2\varepsilon}\left(\frac{N_S}{N_d}\right)^2 = \frac{eN_d}{2\varepsilon}\frac{{N_S}^2}{N_d} \propto \frac{{N_S}^2}{N_d}.$$

This shows that the barrier decreases with increasing doping N_d, as expected, and increases quadratically with the surface states density N_S. Note the crucial distinction between N_S (m^{-2}) and N_{SS} (m^{-2} eV^{-1}).

Another important point is that the depletion width is related to the barrier:

$$x_0 = \sqrt{\frac{2\varepsilon V_S}{eN_d}}.$$

An increased barrier height enlarges the depletion region. In fact, as the surface charge increases, the material must generate a correspondingly larger space charge to compensate for it, which requires the depletion region to extend deeper into the semiconductor.

Let us now examine what occurs when the metal is replaced by a common environmental gas (either reducing or oxidizing). Polycrystalline semiconductor materials used in gas sensing are typically composed of nanostructured elements, among which nanograins (also referred to as nanospheres) are the most prevalent. These sensors are generally synthesized to achieve nanograins that are as uniform as possible, making it feasible to define an average grain size, such as an average radius or diameter. Therefore, we can apply the ideal model to the real system by assuming that all grains are uniform and share the same size, corresponding to the calculated average value. In Fig. 6.30, a layer of nanograins from an ideal sensor in contact with a surface gas is illustrated (in reality, there are multiple layers, the grains are less spherical, and they have slightly different dimensions); across the layer, a potential difference is applied.

Let us now examine the contact between two nanograins to begin formulating the conduction model. Each grain can be represented as a sphere composed of a central bulk region and an outer space-charge region (depleted of free carriers, as in the Schottky model) referred to as the **depletion region**. When a gas interacts with the grain surface, their interaction gives rise to a **double Schottky barrier**, as illustrated in Fig. 6.31.

In this configuration, space charge (qN_d) is present beneath the barrier, surface charge (qN_S) accumulates at the interface, and the bulk remains electrically neutral. In the bulk region, the potential is zero ($\Phi = 0$) far from the junction, which leads to:

$$\rho = qN_d - qN_d e^{\frac{q\Phi}{K_B T}} \rightarrow 0.$$

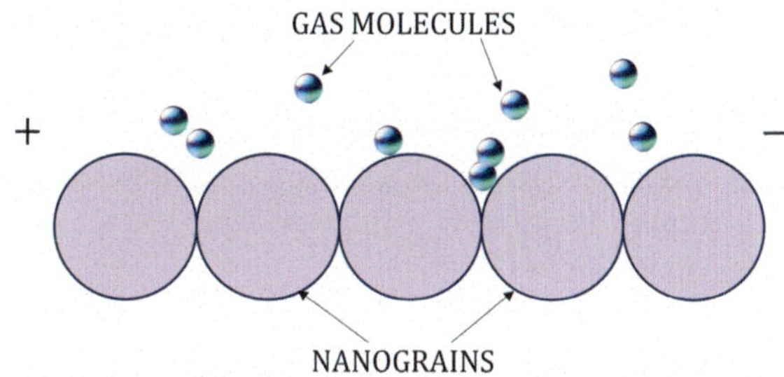

Fig. 6.30 Schematization of an ideal layer of semiconductor nanograins in contact with gas molecules. A potential difference is applied at the edges of the layer

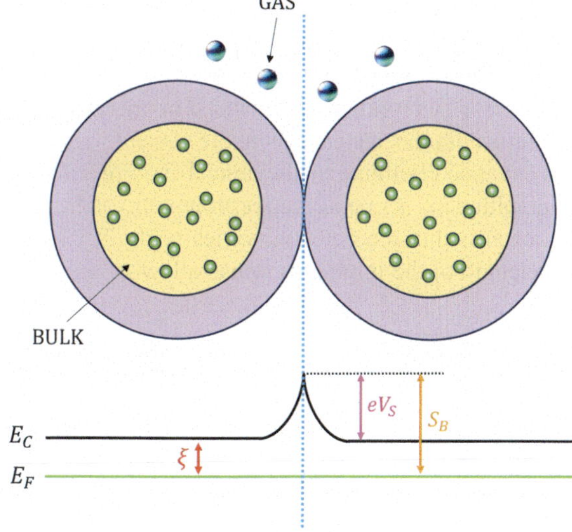

Fig. 6.31 Double Schottky barrier between two ideal semiconductor nanograins. The bulk is represented by the yellow region, where the electrical charge is neutral, and conduction electrons are represented by green dots and the depletion region is represented by the pink shell, in which there are no free charges. The energetic band diagram (with E_C and E_F) is reported below nanograins

Moreover, the electric field within each grain vanishes, meaning that the energy bands are flat throughout the bulk. In summary, each grain behaves as a cluster of monocrystals that together form an approximately spherical polycrystalline nanostructure, whose conduction mechanism will be examined in the subsequent analysis.

Now, focusing on the barrier, the electron density in the bulk region is given by $n_B = N_d$, as the total charge must be neutral. At the interface, however, the electron density becomes:

$$n_S = N_d e^{\frac{q\Phi(\text{surface})}{K_B T}} = N_d e^{\frac{-qV_S}{K_B T}},$$

where n_S refers the surface of the grain. These electrons are responsible for the current flowing from one grain to the next, as they are the only carriers with sufficient energy to overcome the barrier. To derive the expression for the conductance, we recall that the electron current density j_n is given by:

$$j_n = n\mu_n \frac{\partial E_{Fn}}{\partial x},$$

where E_{Fn} is the quasi-Fermi level generated by an applying a potential difference across the two edges of the grain layer, as shown in Fig. 6.32.

The quasi-Fermi level exhibits a slope, and the difference between the two ends of the grain chain is:

$$\frac{\Delta E_{Fn}}{\Delta x} = q\vec{F},$$

where $\vec{F}$ is the average electric field at the junction. Thus, considering a one-dimensional case:

$$j_n = nq\mu_n F.$$

It is therefore clear that, to first order, the current density is proportional to the electric field, with $nq\mu_n$ acting as the proportionality constant. According to Ohm's law, this

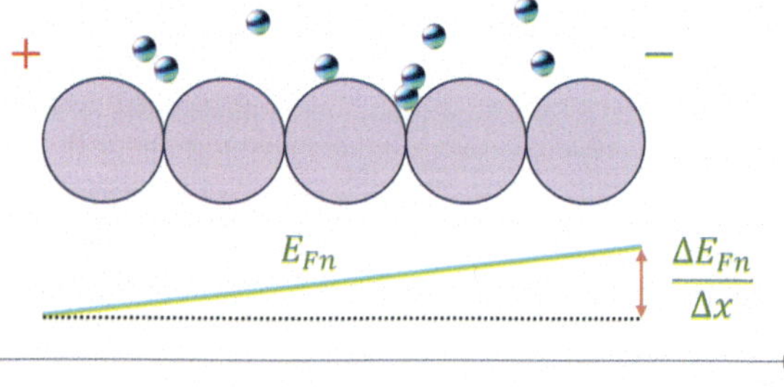

Fig. 6.32 Quasi-fermi level slope in a layer of nanograins with a potential difference applied across the two edges of the grain layer

constant corresponds to the conductivity σ:

$$\sigma = nq\mu_n.$$

Since the relevant electron density is the surface density n_S, the conductivity becomes:

$$\sigma = q\mu_n N_d e^{\frac{-qV_S}{K_B T}},$$

and by defining the pre-exponential factor, known as the **bulk conductivity**:

$$\sigma_0 = q\mu_n N_d,$$

we obtain:

$$\sigma = \sigma_0 e^{\frac{-qV_S}{K_B T}}.$$

To determine the conductance of the material, geometric factors must also be included. The resistance is given by $R = \rho\ l/S$ (where l and S denote the length and cross-sectional area of the material), and thus the conductance is $G = 1/R = \rho^{-1}S/l = \sigma S/l$. Therefore:

$$G = G_0 e^{\frac{-qV_S}{K_B T}},$$

which represents the fundamental relation used to describe the behavior of various types of gas sensors. If the system were pinned (as in covalent semiconductors), the barrier height would remain constant even when the surface gas changes, making sensing impossible. In the following section, we examine how this equation underpins the operating principle of chemoresistive gas sensors.

References

1. Sze SM, Ng KK (2006) Physics of semiconductor devices, 3rd edn. Wiley, Hoboken, NJ
2. Pierret RF (1996) Semiconductor device fundamentals, 2nd edn. Addison-Wesley, Reading, Massachusetts
3. Streetman BG, Banerjee S (2015) Solid state electronic devices, 7th edn., Global Edition. Pearson
4. Bardeen J (1947) Surface states and rectification at a metal semiconductor contact. Phys Rev 71:717–727
5. Rhoderick EH, Williams RH (1988) Metal-semiconductor contacts. Oxford Science Publications, Oxford
6. Madou MJ, Morrison SR (1997) Chemical sensing mechanisms in solid-state devices. In: Chemical sensing with solid state devices. Academic Press, pp 74–83
7. Kurtin S, McGill TC, Mead CA (1969) Fundamental energy levels at metal–semiconductor interfaces. Phys Rev Lett 26:1433–1436

8. Shaw MP (1985) Semiconductor interfaces and contacts. In: Handbook on semiconductors. North-Holland Physics Publishing, Amsterdam, pp 51–60
9. Flores F (1995) Electronic structure at surfaces and interfaces. Surf Rev Lett 2:513–520
10. Brillson LJ (1992) Metal/semiconductor interfaces. In: Moss TS (ed) Handbook of semiconductors. Elsevier, Amsterdam
11. Stilies K, Khan A (1988) Surface electronic states at semiconductor interfaces. Phys Rev Lett 60:440–443
12. Seto JYW (1975) The electrical properties of polycrystalline silicon films. J Appl Phys 46:5247–5254
13. Hasegawa H, Sato T, Kasai S (2000) Surface and interface properties of compound semiconductors. Appl Surf Sci 166:92–100
14. Mikhelashvili V, Eisenstein G, Uzdin R (2001) Grain boundary effects in polycrystalline semiconductors. Solid-State Electron 45:143–150
15. Aldao CM, Malagù C (2012) Electrical transport and surface states in metal oxide gas sensors. J Appl Phys 112:024518
16. Malagù C, Carotta MC, Guidi V, Martinelli G (2004) Unpinning of Fermi level in nanocrystalline semiconductors. Appl Phys Lett 84:4158–4160
17. Malagù C, Carotta MC, Comini E, Faglia G, Giberti A, Guidi V, Maffeis TGG, Martinelli G, Sberveglieri G, Wilks SP (2005) Photo-induced unpinning of Fermi level in WO3. Sensors 5:594–603
18. Malagù C, Martinelli G, Ponce MA, Aldao CM (2008) Unpinning of the Fermi level and tunneling in metal oxide semiconductors. Appl Phys Lett 92:162104
19. Guidi V, Malagù C, Martinelli G, Cozzolino S, Padula M (2006) Semiclassical approach for determination of the intergranular energy barrier height in very-fine nanograins. Appl Phys Lett 89:203105
20. Malagù C, Carotta MC, Giberti A, Guidi V, Martinelli G, Ponce MA, Castro MS, Aldao CM (2009) Two mechanisms of conduction in polycrystalline SnO2. Sens Actuators B Chem 136:230–234
21. Guidi V, Carotta MC, Malagù C, Martinelli G (2009) Modeling of the inter-granular energy-barrier height in very-fine nanograins through a semi-classical approach. Sens Actuators B Chem 137:521–523
22. Malagù C, Carotta MC, Martinelli G, Ponce MA, Castro MS, Aldao CM (2009) Field-assisted and thermionic contributions to conductance in SnO2 thick-films. J Sensors 2009:402527
23. Ponce MA, Ramírez MA, Parra R, Malagù C, Castro MS, Bueno PR, Varela JA (2010) Influence of degradation on the electrical conduction process in ZnO and SnO2-based varistors. J Appl Phys 108:074505
24. Vaz ICF, Macchi CE, Somoza A, Rocha LSR, Longo E, Cabral L, da Silva EZ, Simões AZ, Zonta G, Malagù C, Desimone PM, Ponce MA, Moura F (2022) Electrical transport mechanisms of neodymium-doped rare-earth semiconductors. J Mater Sci Mater Electron 33(11632–11):649
25. Martinelli G, Carotta MC, Malagù C (2000) Charge transport and surface effects in metal oxide sensors. Electron Technol 33:40–46
26. Carotta MC, Benetti M, Ferrari E, Giberti A, Malagù C, Nagliati M, Vendemiati B, Martinelli G (2007) Basic interpretation of thick film gas sensors for atmospheric application. Sens Actuators B 126:672–677
27. Solis JL, Hoel A, Lantto V, Granqvist CG (2001) Gas-sensing properties of nanocrystalline metal oxides. J Appl Phys 89:2727–2732
28. Diederich L, Aebi P, Kuttel OM, Schlapbach L (1999) Surface electronic structure studied by photoemission. Surf Sci 424:L314–L320
29. Spicer WE, Chye PW, Skeath PR, Su CY, Lindau I (1979) Unified defect model and the interface states. J Vac Sci Technol 16:1422–1433
30. Rantala TT, Rantala TS, Lantto V (1999) Electronic structure of SnO_2 surfaces. Surf Sci 420:103–109
31. Orton JW, Powell M (1980) The Hall effect in polycrystalline and powdered semiconductors. Rep Prog Phys 43:81–119

32. McAleer JF, Moseley PT, Norris JOW, Williams DE (1987) Tin dioxide gas sensors: dopant and defect chemistry. J Chem Soc Faraday Trans 1(83):1323–1346
33. Martinelli G, Carotta MC (1995) Thick-film gas sensors. Sens Actuators B Chem 23:157–161

Chapter 7
Metal Oxide (MOX) Gas Sensors

There is plenty of room at the bottom.
Richard P. Feynman

Metal oxide (MOX) gas sensors represent one of the most established and versatile classes of chemical sensors, widely employed for the detection of a broad range of gaseous species. Their operation relies primarily on chemoresistive effects, where changes in surface interactions between target molecules and the sensing layer are translated into measurable variations in electrical resistance. A deeper understanding of these devices requires linking their macroscopic behavior to microscopic and nanoscale phenomena, including charge transport, potential barriers at grain boundaries, and the role of intrinsic defects such as oxygen vacancies. To this end, both experimental techniques and theoretical models are employed, ranging from Arrhenius-plot analyses of temperature-dependent conductivity to tunneling-based approaches and scanning probe methods. Particular attention is devoted to the influence of electronic structure and defect states on sensing performance, as well as to the application of semiconductor approximations in the description of nanograins. By combining these perspectives, this Chapter provides a comprehensive framework for interpreting the physical mechanisms that govern MOX gas sensor operation and highlights their relevance for advancing sensing technologies. This Chapter has been developed through an integrated elaboration of multiple sources cited herein [1–33], together with the authors' original contribution, with selected references being explicitly discussed at specific points in the text.

7.1 Chemoresistive Gas Sensors

A gas sensor is a device in which a measurable physical property undergoes a change when the chemical or physical composition of the surrounding atmosphere is modified. More specifically, a chemical sensor responds to the presence of specific chemical species through a variation in a physical signal (typically optical, electrical, or electrochemical) arising from physicochemical interactions at its surface or within its

© The Author(s), under exclusive license to Springer Nature Switzerland AG 2026

C. Malagù and G. Zonta, *Gas Sensors*, https://doi.org/10.1007/978-3-032-21614-4_7

active layer. Among chemical sensors, **chemoresistive gas sensors** are distinguished by the fact that their electrical conductance changes as a consequence of surface reactions or adsorption processes involving ambient gas molecules, a behavior referred to as chemoresistivity.

Let us focus on the two prototypical metal-oxide sensing materials, SnO_2 and TiO_2. Both compounds are wide-bandgap n-type semiconductors, with bandgap energies of approximately 3.7 eV and 3.1 eV, respectively, placing them near the boundary with insulating materials, which generally exhibit bandgaps exceeding ~4 eV. Their n-type character originates from oxygen vacancies, which act as shallow donor states located close to the conduction-band edge (E_C). Reported donor-level energies include ~0.03 eV and 0.15 eV for SnO_2 (Samson et al. 1973) and ~0.7 eV for TiO_2 [1, 2]. These intrinsic defects, unavoidably introduced during synthesis, provide the free carriers responsible for electrical conduction. In the absence of oxygen vacancies, both SnO_2 and TiO_2 would exhibit insulating behavior. Figure 7.1 shows the unit cells of the two oxides in their rutile structures, highlighting the close similarity of their lattice frameworks.

In Table 7.1 some main characteristics of these two materials are reported.

As previously discussed, oxygen vacancies constitute stoichiometric defects in MOX materials and act as shallow donor levels. These vacancies arise from the absence of an oxygen atom at a specific lattice site (Fig. 7.2). They are introduced during material synthesis and subsequent annealing treatments (see Chap. 8).

These defects play a central role in the formation of the surface barrier eV_S when surface acceptor states (such as ionosorbed oxygen species) become charged. When

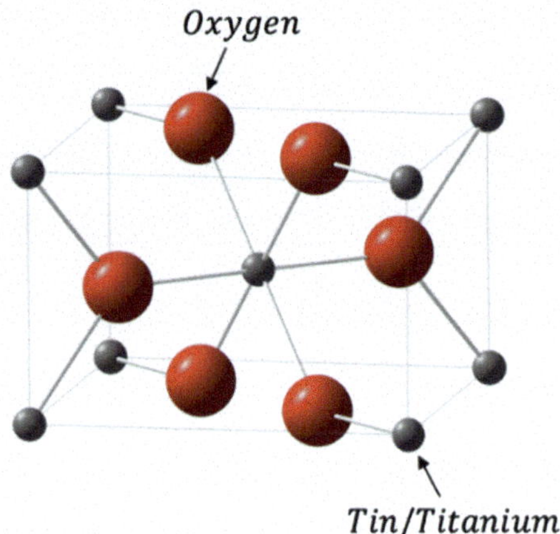

Fig. 7.1 Tetragonal unit cell of SnO_2 and TiO_2 in their rutile structures

Table 7.1 Some of the main characteristics of SnO_2 and TiO_2

Material	Unit cell	Space group	Ionic radius	a	c
SnO_2	Tetragonal	$D_{4h}^{14} - P4_2/nmm$	0.71 Å (of Sn^{4+})	4.737 Å	3.186 Å
TiO_2	Tetragonal	$D_{4h}^{14} - P4_2/nmm$	0.68 Å (of Ti^{4+})	4.594 Å	2.958 Å

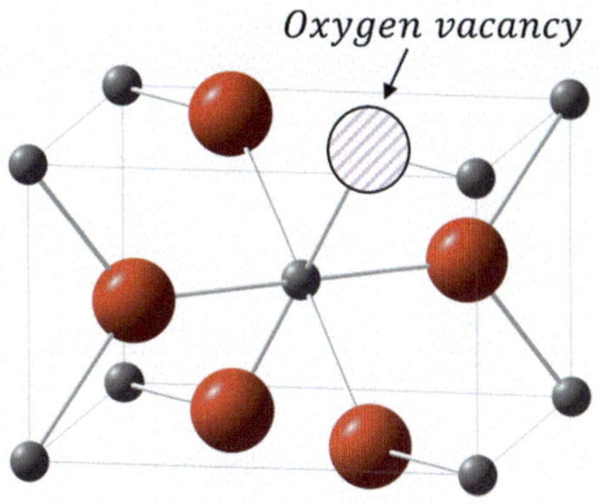

Fig. 7.2 Oxygen vacancy as intrinsic defect responsible for conduction properties of SnO_2 and TiO_2

an oxygen vacancy is generated at a lattice site (approximately one cell per million), and the doping level is sufficiently high (~$10^{17}-10^{18}$ cm^{-3}), the material exhibits semiconducting behavior. This intrinsic doping gives rise to donor states that are fully ionized at temperatures near 300 °C, allowing us to consider $N_d = N_d^+$ and to apply the previously developed formalism.

Let us now analyze the barrier formation process, based on Bardeen's model but approached from a different perspective by examining the energy band diagram in Fig. 7.3, which represents a double-layer configuration.

Even in the absence of impurities, the surface constitutes an intrinsic imperfection because the periodicity of the crystal is interrupted. It can therefore be assumed that both donor and acceptor states exist at the surface, resulting in a level Φ_0 that equilibrates it. One may further consider the existence of an electrochemical potential

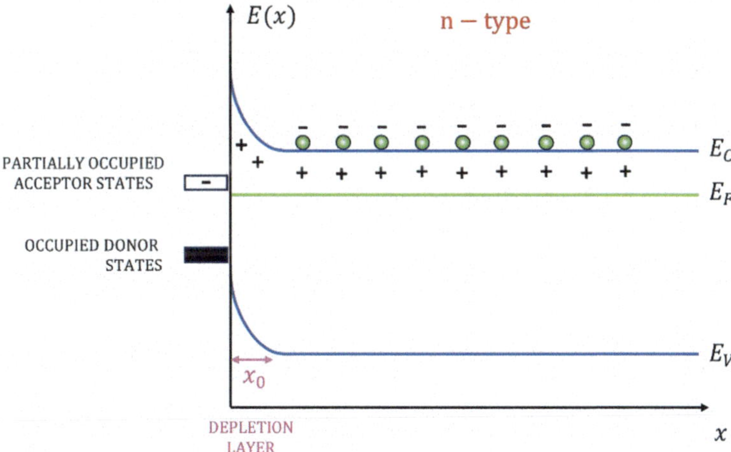

Fig. 7.3 Double-layer model. Electrons from the conduction band are captured by surface states, resulting in a negatively charged surface and positively charged donors near the surface

associated with these surface states such that, at equilibrium, the electrochemical potential (and thus the Fermi level) is uniform throughout the semiconductor. This interpretation aligns with the electrochemical model and mirrors Bardeen's approach, in which Φ_0 governs the Fermi-level position. In a pinned semiconductor, $E_F = \Phi_0$; in an unpinned system, the Fermi level may differ from Φ_0, allowing variations in the surface barrier S_B in response to changes in the ambient gas composition, as exploited in gas-sensing applications. Referring to the two-grain representation in Fig. 7.4, we can now describe the gas-sensing mechanism.

Consider an n-type semiconductor. It is well established that sensing layers are polycrystalline films composed of metal oxide nanoparticles capable of reversibly adsorbing gas-phase species. Oxygen from the surrounding environment is chemisorbed onto grain surfaces, giving rise to a depletion layer. Due to its high electronegativity (greater than that of the semiconductor itself) oxygen attracts electrons from the conduction band, forming O^- or O_2^- species that populate surface states. This electron transfer produces a region depleted of free carriers and leads to the development of a potential barrier.

In a n-type semiconductor, these surface states behave as ionized acceptors that compensate for the positive charge of the space-charge region. Figure 7.5 presents transmission electron microscopy (TEM) images illustrating the granular

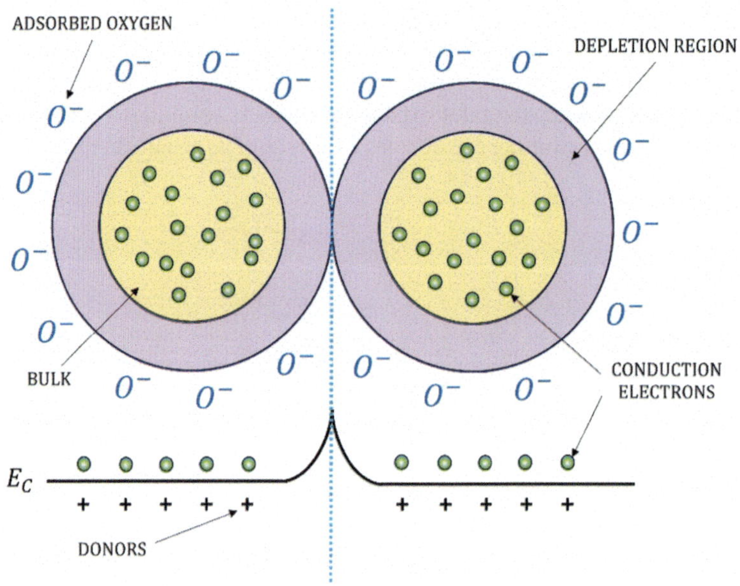

Fig. 7.4 Formation of the depletion layer in semiconductor nanograins. The bulk is represented by the yellow region, where the electrical charge is neutral, and conduction electrons are represented by green dots and the depletion region is represented by the pink shell, in which there are no free charges. The energetic band diagram with the double barrier is reported below nanograins

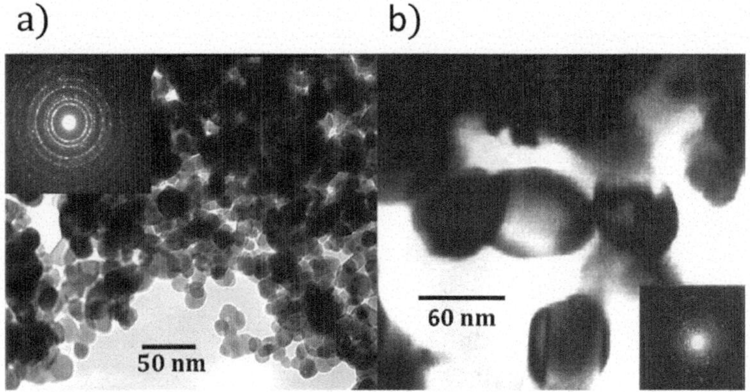

Fig. 7.5 TEM images of a SnO_2 film at two different magnifications

morphology of a SnO_2 film. In particular, Fig. 7.5b shows three contiguous grains aligned in a row.

In Sect. 7.2, we will examine how the conduction mechanisms of chemoresistive sensors, as described here, can be applied to derive the Arrhenius plot.

7.2 Arrhenius Plot

The **Arrhenius plot** is a well-established method for determining the activation energy through the "temperature jumps" technique. The conduction law exhibits an exponential dependence on the inverse of temperature:

$$k = ae^{-\frac{E_A}{k_B T}}$$

where a is the **pre-exponential factor** and E_A is the **activation energy**. Physical laws exhibiting this functional dependence can be plotted using the Arrhenius plot, where the observable is shown on the y-axis (in logarithmic scale), and the inverse of the temperature $(1/T)$ is plotted on the x-axis:

$$\ln k = \ln a + m\frac{1}{T}$$

with $m = -E_A/k_B$.

Since the conductance of a chemoresistive gas sensor is described by:

$$G = G_0 e^{\frac{-qV_S}{K_B T}},$$

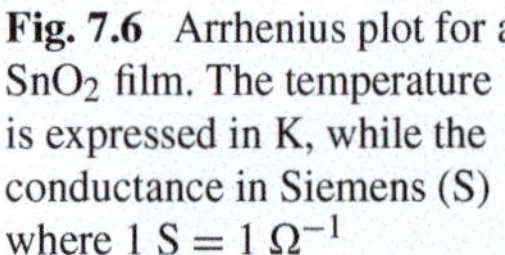

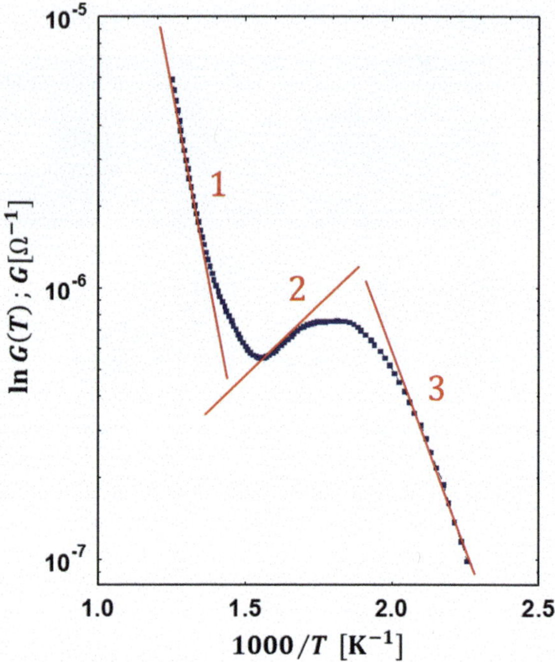

Fig. 7.6 Arrhenius plot for a SnO_2 film. The temperature is expressed in K, while the conductance in Siemens (S) where $1\ S = 1\ \Omega^{-1}$

Figure 7.6 presents the Arrhenius plot for a SnO_2 film, typically fabricated for gas sensing applications.

From Fig. 7.6, it is evident that the curve features three distinct slopes, labeled 1, 2 and 3, in order of increasing temperature (from right to left). Slope 1 corresponds to an initial increase in conductance with temperature, while slope 3 exhibits a similar trend at higher temperatures. Slope 2 represents an intermediate temperature interval (between approximately points 1.5 and 2 in Fig. 7.6), in which the conductance approaches a plateau. Figure 7.7 presents experimental results showing the evolution of conductance with temperature from the work of Chang et al. and Lantto et al. [3, 4].

(In Fig. 7.7a, the Arrhenius plot for SnO_2 in dry air clearly displays a central plateau-like region. The prevailing interpretation in the literature is that chemisorbed molecular oxygen undergoes a transition within the temperature range from 150 – 200 °C up to approximately 300 – 400 °C. At low temperatures, surface states originate primarily from adsorbed molecular oxygen, which captures electrons and induces additional band bending. As temperature increases, the molecular oxygen dissociates into atomic oxygen, which exhibits greater electronegativity than the molecular species. Consequently, surface states form even more easily, as atomic oxygen is more reactive. Since the barrier height increases proportionally with the square of the surface-state density N_S, an increase in these states enhances the barrier height, thereby flattening the conductance curve. This phenomenon is clearly illustrated in Fig. 7.7b.

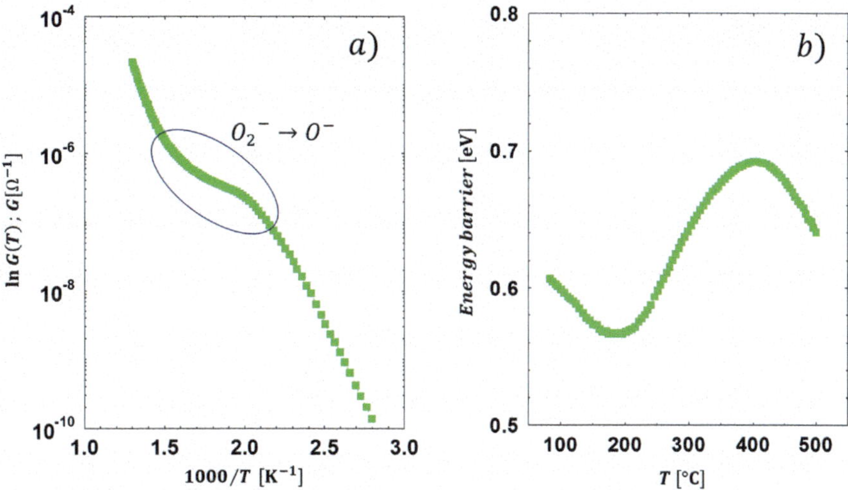

Fig. 7.7 **a** Arrhenius plot for a SnO_2 film in dry air the central region corresponding to the $O_2^- \rightarrow O^-$ transition is circled; **b** temperature dependence of the surface potential barrier for a SnO_2 film in air

Thus, two opposing processes occur by increasing temperature: (1) an increase in conductance due to thermal activation, and (2) $O_2^- \rightarrow O^-$ transformation, which suppresses the conductance. Once all molecular oxygen has been converted to atomic oxygen, the conductance resumes its increase, resulting in the steeper slope observed at higher temperatures.

(This trend is absent in TiO_2. From the experimental data shown in Fig. 7.8a, the Arrhenius plot for TiO_2 (blue curve) displays only two slopes in contrast with SnO_2 (green curve). Thus, the $O_2^- \rightarrow O^-$ transition does not occur on TiO_2 surfaces. The conductance of TiO_2 increases only weakly at low temperatures, but then rises exponentially with a sudden change in slope as temperature continues to increase.

To extract the barrier height directly from experimental data, the "**temperature-jump**" method (also known as stimulated temperature conductance measurement) can be employed [5, 6].

The method relies on inducing rapid stimulated temperature variations such that the barrier height may be considered constant during the transition, while the conductance changes solely as a result of the variation in free carriers. In practice, a series of few conductance measurements is performed using a digital multimeter, applying abrupt temperature changes (hence the term "jump") from an initial temperature T_i to a selected value (see Fig. 7.9, where $T_i = 250$ °C). Before initiating the measurement (at $t = 0$), the conductance must be allowed to reach a stable baseline corresponding to thermal equilibrium between the sensor and the environment. In Fig. 7.9, the first temperature jump corresponds to $\Delta T = 30$ °C (from $T_i = 250$ °C to $T_{f1} = 280$ °C, as illustrated by the green and blue curves). After this step, the conductance is allowed to stabilize at a new plateau before the system is returned to T_i (baseline).

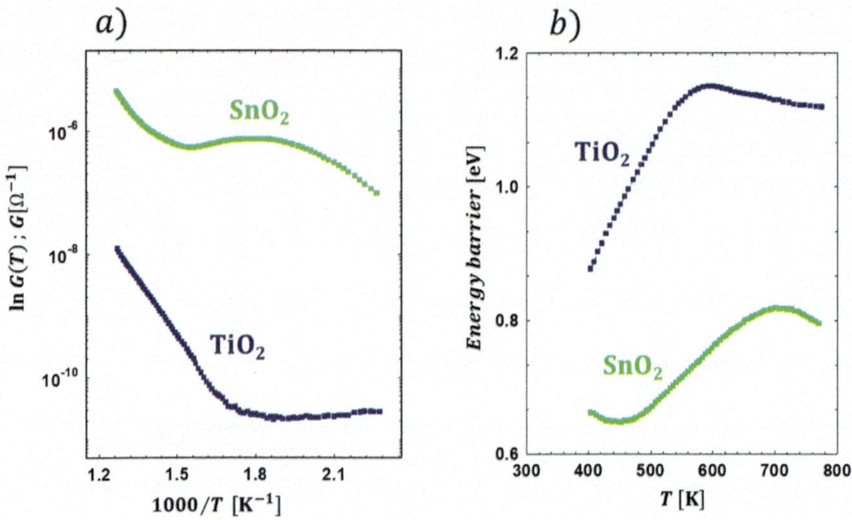

Fig. 7.8 **a** Conductance and **b** surface barrier evolution with temperature for SnO_2 and TiO_2

The procedure is then repeated with larger and smaller final temperatures (e.g., $T_{f2} = 320$ °C and $T_{f3} = 220$ °C). Only three distinct final temperatures are required for the analysis.

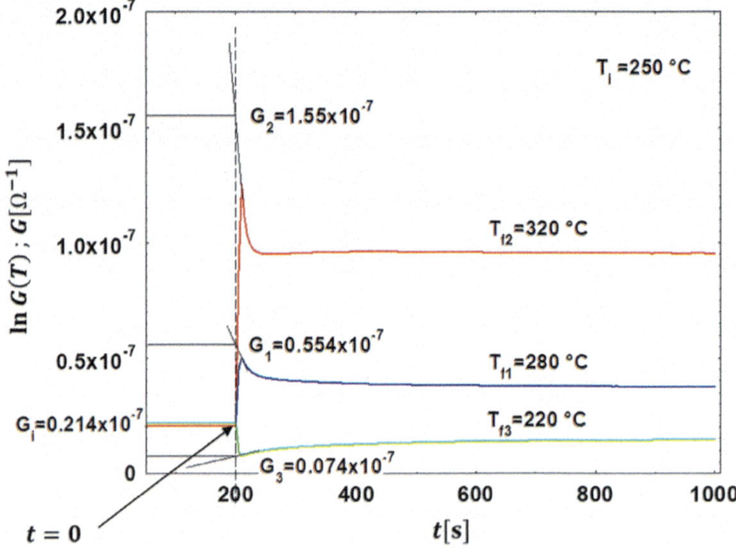

Fig. 7.9 Evaluation of the potential-barrier height qV_S by stimulated temperature conductance measurements

To interpret the data, the general conductance expression is applied to both the initial value G_i and each final value G_n:

$$G_i = G_0 e^{-\frac{qV_S(T_i)}{k_B T_i}},$$

$$G_n = G_0 e^{-\frac{qV_S(T_i)}{k_B T_{fn}}},$$

where G_0 is the temperature-independent pre-exponential factor, and the barrier height in the numerator corresponds to the initial temperature T_i. Indeed, the conductance values extrapolated at $t = 0$ (by extending the peak of the transient response) correspond to the new temperature T_{fn}, but retain the barrier height associated with the initial temperature T_i, i.e., $qV_S(T_i)$. Only in the second phase of the measurement (plateau) does the barrier height correspond to the final temperature T_{fn}, i.e., $qV_S(T_{fn})$.

Repeating the extrapolation for the three temperature jumps yields three values G_n, which can be plotted to form a straight line described by:

$$\ln \frac{G_n}{G_i} = -\frac{qV_S}{k_B}\left(\frac{1}{T_{fn}} - \frac{1}{T_i}\right).$$

Because this relation is linear, it can be fitted using the least-squares method (Fig. 7.10), providing the coefficients of the best linear fit.

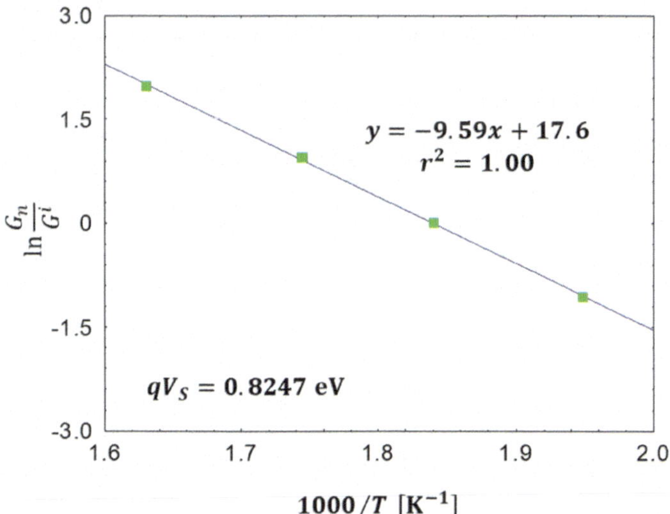

Fig. 7.10 Plot of ln G_n/G_i as a function of 1000/T [K^{-1}]. The green points represent the experimental data, and the best fitting expression y= −9.59x+17.6 (where the two coefficients correspond to the slope and intercept) yields r^2=1.00

Multiplying the angular coefficient of the fit of $\ln(G_n/G_i)$ by k_B provides the value of the potential barrier at the corresponding temperature. Assuming that G_0 is nearly temperature-independent, this barrier height can be substituted into the inverse expression for G_i, allowing G_0 to be determined as:

$$G_0 = G_i e^{\frac{qV_S(T_i)}{k_B T_i}}.$$

From this, the temperature dependence of qV_S can be derived, using $\ln G$ as a function of temperature obtained from the Arrhenius plot:

$$qV_S = k_B T(\ln G - \ln G_0).$$

This relation leads to the curve presented in Fig. 7.7b. Two limitations of the method should be noted: (i) experimental uncertainties inherent to the operator, and (ii) the assumption that only the thermionic contribution to the barrier is relevant (neglecting tunneling), which causes the barrier height to be underestimated by approximately 10%. Despite these limitations, this remains the most reliable experimental technique for determining the barrier height. Figure 7.11 shows the temperature dependence of the energy barrier (in dry air) for films composed of different materials (fired at 850 ° C).

In all samples containing Ti, the energy barriers are characteristic of TiO_2 sensors, displaying only two distinct barrier-energy regions, which are significantly higher

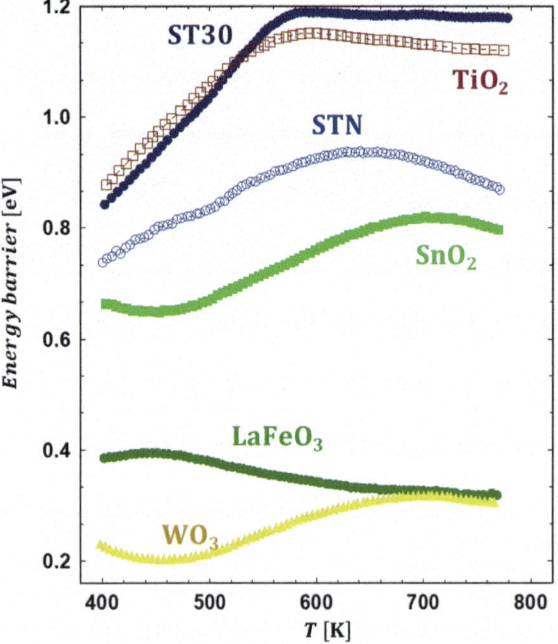

Fig. 7.11 Example of energy barrier qV_S [eV] as a function of the temperature T [K] calculated for different sensing materials

than those of SnO_2 sensors. Unlike SnO_2, the sensor TiO_2, as well as ST30 (mixture of tin oxide and titanium oxide at 30%) and STN (mixture of tin, titanium, and niobium oxides), do not exhibit a minimum around 200 ° C, a feature associated with of the onset of the $O_2^- \rightarrow O^-$ transformation at SnO_2 grain boundaries. WO_3 behaves similarly to SnO_2, but with a much lower energy barrier, likely due to the exclusive adsorption of $O_2{}^-$ ions. $LaFeO_3$ clearly exhibits the characteristics of a p-type semiconductor, with a comparatively low energy barrier.

Overall, the temperature range in which the Schottky barrier varies extends from approximately 400 K to 750–780 K. For SnO_2, the barrier height spans 0.6–0.8 eV, and in general, the Schottky barrier height does not exceed 1.1–1.2 eV, even when accounting for experimental uncertainties.

7.3 DA Applied to Nanograins

Let us now reconsider the geometric situation of two nanograins, illustrated in Fig. 7.12.

For this analysis, the grain radius is denoted by R, the bulk radius by R_0, the depletion width by Λ, and the general radial coordinate by r. From the previous derivation, the potential V_S is given by:

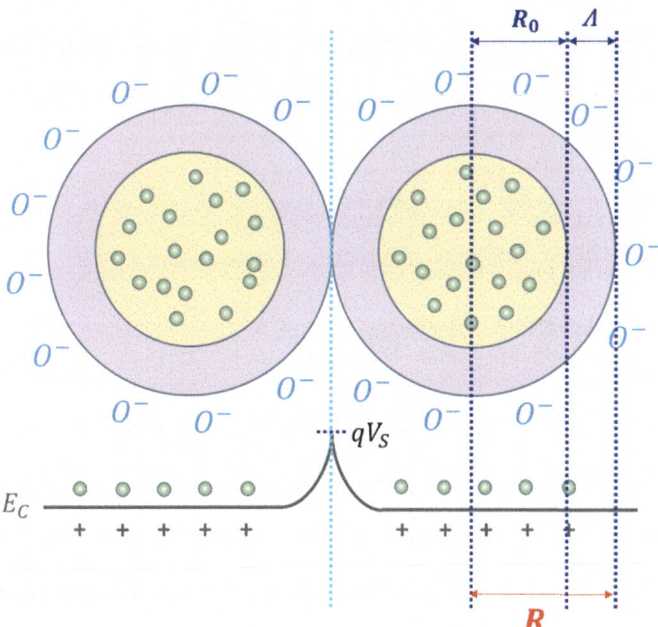

Fig. 7.12 Two nanograins, indicating the barrier height qV_S, the grain radius R, the bulk radius R_0, and the depletion width Λ. The radial coordinate is denoted by r

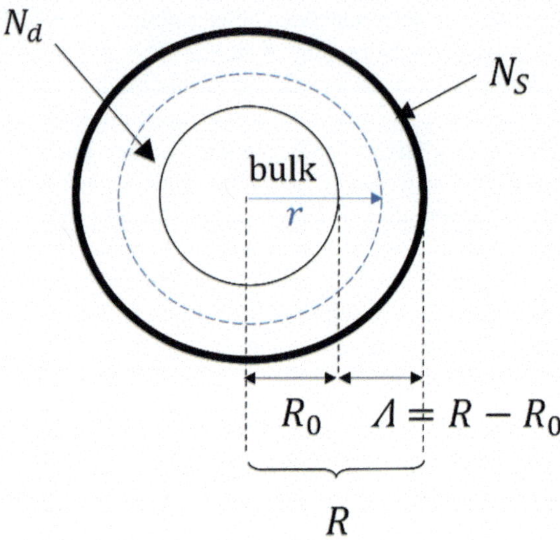

Fig. 7.13 Spherical grain representation: the grain radius is R, the bulk radius is R_0, the depletion width is Λ, and r is the radial coordinate. N_d and N_S denote the densities of donors and surface states, respectively

$$V_S = \frac{qN_d}{2\varepsilon_0\varepsilon_r}\Lambda^2.$$

When $R \approx \Lambda$, the two barriers begin to overlap at the grain center. In this regime, the classical Schottky expression no longer holds, and corrections proportional to Λ/R must be introduced, as discussed below. For simplicity, a spherical grain geometry is assumed (Fig. 7.13), though the specific structure is not crucial.

Considering now the Laplacian in spherical coordinates:

$$\Delta\Phi = \frac{1}{r}\frac{\partial^2}{\partial r^2}(r\Phi) + \frac{1}{r^2 \sin\theta}\frac{\partial}{\partial\theta}\left(\sin\theta\frac{\partial\Phi}{\partial\theta}\right) + \frac{1}{r^2\sin^2\theta}\frac{\partial^2\Phi}{\partial\varphi^2},$$

the spherical symmetry of the system reduces Poisson's equation to:

$$\frac{1}{r}\frac{\partial^2}{\partial r^2}[r\Phi(\mathrm{r})] = -\frac{qN_d}{\varepsilon},$$

in fact, only the radial term of the Laplacian differs from zero. To obtain a unique solution, we must impose that the electric field is zero in the bulk (flat band region) as well as the potential (by convention):

$$-\left.\frac{\partial\Phi(r)}{\partial r}\right|_{r=R_0} = 0, \quad \Phi(R_0) = 0.$$

At $r = R_0$, we proceed analogously to the D.A, beginning with:

$$\frac{d^2}{dr^2}[r\Phi(\mathrm{r})] = -ar,$$

where $a = qN_d/\varepsilon$. Integrating twice:

$$\frac{d}{dr}[r\Phi(\mathrm{r})] = -\frac{ar^2}{2} + c_1,$$

$$r\Phi(\mathrm{r}) = -\frac{ar^3}{6} + rc_1 + c_2.$$

and applying the boundary conditions yields:

$$\Phi(R_0) = -\frac{aR_0{}^2}{6} + c_1 + \frac{c_2}{R_0} = 0,$$

$$\Phi'(R_0) = -\frac{aR_0}{3} - \frac{c_2}{R_0{}^2} = 0,$$

from which:

$$c_1 = \frac{aR_0{}^2}{2}, \quad c_2 = -\frac{aR_0{}^3}{3}.$$

Thus, the potential becomes:

$$\Phi(r) = -\frac{qN_d}{6\varepsilon}r^2 - \frac{qN_d}{3\varepsilon}\frac{R_0{}^3}{r} + \frac{qN_d}{2\varepsilon}R_0{}^2.$$

This expression shows that the potential contains both a quadratic term in r and a term that varies with $1/r$. Evaluating the potential at the grain surface, we set $\Phi(r = R) = -V_S$, where V_S is experimentally determined (e.g., via the temperature-jump method). Hence:

$$V_S = \frac{qN_d}{6\varepsilon}R^2 + \frac{qN_d}{3\varepsilon}\frac{R_0{}^3}{R} - \frac{qN_d}{2\varepsilon}R_0{}^2,$$

which, using $R_0 = R - \Lambda$ becomes:

$$V_S = \frac{qN_d}{6\varepsilon}R^2 + \frac{qN_d}{3\varepsilon}\frac{(R-\Lambda)^3}{R} - \frac{qN_d}{2\varepsilon}(R-\Lambda)^2,$$

and simplifying further:

$$V_S = \frac{qN_d}{6\varepsilon}\left(3\Lambda^2 - \frac{2\Lambda^3}{R}\right) = \frac{qN_d}{2\varepsilon}\Lambda^2\left(1 - \frac{2\Lambda}{3R}\right).$$

Charge neutrality for a single grain requires:

$$qN_d\left(\frac{4}{3}\pi R^3 - \frac{4}{3}\pi {R_0}^3\right) = qN_S 4\pi R^2.$$

representing the equality between the total charge in the depletion shell (on the left-hand side) and the surface charge (on the right-hand side). The expression in parentheses is the volume of the depletion shell (the difference between the total grain volume and its neutral bulk). Solving for N_S, recalling $\Lambda = R - R_0$:

$$\begin{aligned} N_S &= \tfrac{N_d}{3R^2}\left(R^3 - {R_0}^3\right) \\ &= \tfrac{N_d}{3R^2}\Lambda\left[R^2 + R(R-\Lambda) + (R-\Lambda)^2\right] \\ &= \tfrac{N_d}{3R^2}\Lambda\left[R^2 + R^2 - R\Lambda + R^2 + \Lambda^2 - 2R\Lambda\right] \\ &= N_d\Lambda\left(1 - \tfrac{\Lambda}{R} + \tfrac{\Lambda^2}{3R^2}\right). \end{aligned}$$

In summary, the core of the spherical model is governed by the following two equations:

$$\begin{cases} N_S = N_d\Lambda\left(1 - \frac{\Lambda}{R} + \frac{\Lambda^2}{3R^2}\right) \\ V_S = \frac{qN_d}{2\varepsilon}\Lambda^2\left(1 - \frac{2\Lambda}{3R}\right). \end{cases}$$

Thus, the surface potential V_S is expressed as the classical planar term $(qN_d/2\varepsilon)\Lambda^2$ with the addition of a correction term $-(qN_d/3\varepsilon)(\Lambda^3/R)$, which accounts for curvature effects.

When $\Lambda \ll R$, meaning the bulk is approximately large as the grain itself, the shell becomes nearly planar, and the model reduces to the planar case.

For N_S, two correction terms (first and second order in Λ/R) appear, which collapse into the planar limit (where $N_S = x_o N_d$ in the junction) when $\Lambda \ll R$, as calculated by Malagù et al. [7]. These correction terms become significant when $\Lambda \to R$, and curvature effects start to be relevant.

In the following analysis, N_S and V_S have been evaluated for thick films SnO_2 [8] and TiO_2 [9] at 400 °C in dry air, with R ranging from 50 to 150 nm [8, 9]. Figure 7.14 shows the two unknowns, Λ and N_S, plotted as a function of the mean grain radius R for both materials. All other parameters are known: ε is material-specific; R can be measured from SEM images; V_S is obtained from the Arrhenius plot; and N_d is taken from the literature.

To experimentally carry out these measurements, the grain radius is treated as a variable parameter (plotted on the x-axis) by fabricating sensing materials (both SnO_2 and TiO_2) with different grain sizes. Using the spherical model, the two unknowns are then calculated. From Fig. 7.14a, it is evident that for SnO_2 (synthesized in our laboratory), the surface state density remains constant as R increases from 50 to 150 nm; correspondingly, Λ also remains constant, as expected. Thus, curvature effects are negligible, since $\Lambda \approx 15$ nm $\ll R$.

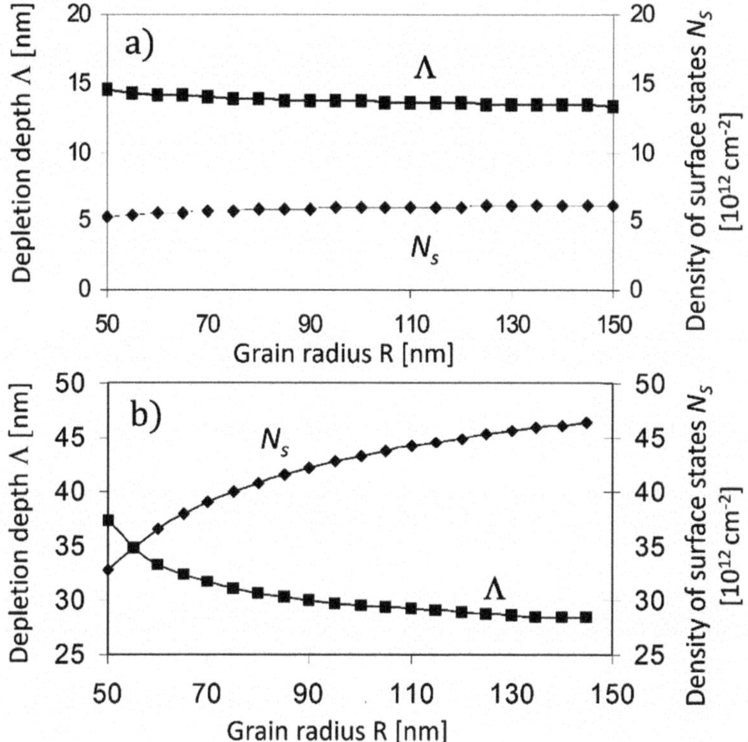

Fig. 7.14 Depletion width Λ and surface state density N_S as a function of mean grain radius R for: **a** SnO_2 ($\varepsilon = 10^{-10}$ F/m; $N_d = 5 \times 10^{18}$ cm^{-3}; $V_S = 0.68$ V); **b** TiO_2 ($\varepsilon = 10^{-9}$ F/m; $N_d = 2 \times 10^{19}$ cm^{-3}; $V_S = 1.12$ V)

In contrast, Fig. 7.14b shows that for TiO_2, Λ may reach 35 − 40 nm, which becomes comparable to the grain radius. In this situation, the entire grain can become depleted, and curvature effects become significant. Moreover, the surface state density decreases as Λ increases.

It is well established that the barrier height depends quadratically on N_S: a high density of surface states corresponds to a high potential barrier. When low concentrations of reducing gases (in the ppm or ppb range) are introduced, they remove chemisorbed oxygen from the sensing surface. If N_S is large, this removal is analogous to taking a glass of water from a swimming pool: the water level remains essentially unchanged. However, when N_S is reduced to the critical condition in which the grain becomes fully depleted, removing only a small number of oxygen species produces a much more appreciable variation, like taking a glass of water from a bucket, which leads to a visible change in water level. Hence, to first order, a reduced value of N_S increases the sensitivity of the sensor to gas variations (for unpinned materials).

For the limiting case $\Lambda = R$, the spherical model gives:

$$\begin{cases} V_S = \frac{1}{3}\left(\frac{qN_d}{2\varepsilon}\Lambda^2\right), \\ N_S = \frac{1}{3}(N_d\Lambda). \end{cases}$$

Both quantities therefore reduce to one-third of their planar-geometry values:

$$\begin{cases} V_S = \frac{1}{3}V_S^{\text{planar}}, \\ N_S = \frac{1}{3}N_S^{\text{planar}}. \end{cases}$$

Let us now consider three TiO_2 materials (fired at different temperatures) produced in the Sensor Laboratory of UNIFE, each characterized by different but highly homogeneous grain sizes, as shown in Fig. 7.15. It was found that firing temperatures above 750 °C induce a phase transition. At 650 °C, the material is in the cassiterite phase; further heating triggers transformation into rutile, which possesses a smaller lattice parameter and "packs" more efficiently, marking the onset of grain coalescence. Unlike SnO_2, which does not exhibit a comparably well-defined nanostructure, TiO_2 is well suited to grain-radius modulation.

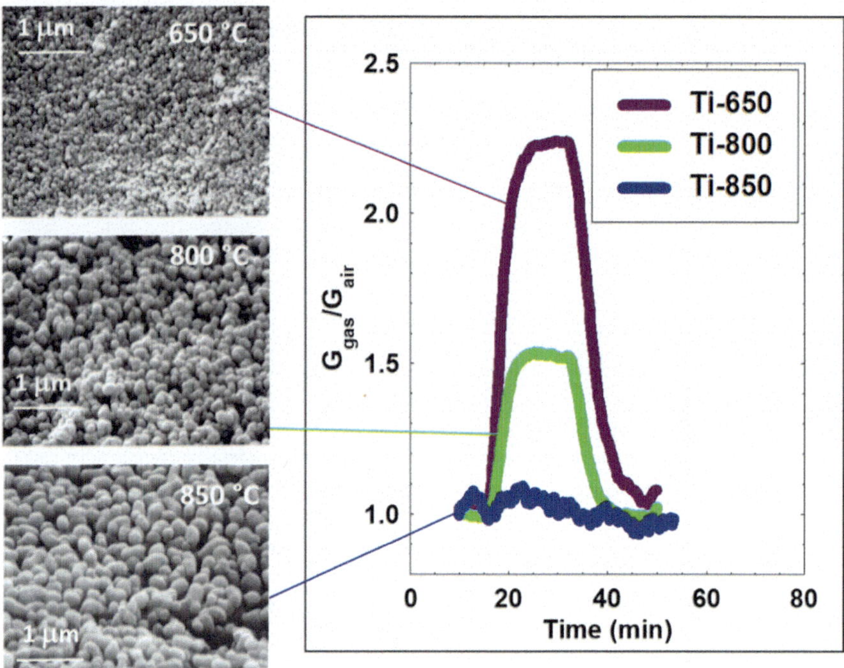

Fig. 7.15 Responses $R = G_{gas}/G_{air}$ of three TiO_2-based gas sensors fired at 650, 800, and 850 °C. G_{gas} is the conductance in the presence of the target gas and G_{air} the baseline conductance in synthetic dry airflow. The response strongly depends on the grain dimensions, which are related to the firing temperature of the sensing material

After extensive measurements, the material fired at 650 ° C, with a grain radius equal to its depletion width (30 − 40 nm), was found to yield optimal sensing responses. Increasing the grain radius (along with the firing temperature) leads to a progressive decrease in response. When the grain radius becomes four to five times larger than the depletion width, the response approaches zero, as experimentally confirmed and shown by the blue curve in Fig. 7.15. This pronounced dependence of response on grain radius is also connected to the unpinning phenomenon, which can be experimentally confirmed using Scanning Tunneling Spectroscopy (STS).

When the grain radius approaches the depletion width, not only does N_S decrease, but a "cleaning" of the bandgap region also occurs, meaning a reduction in the density of states per unit energy (fewer defects). The microscopic origin of this effect is not fully understood, although it is consistently observed at the phenomenological level.

Let us now examine whether a comparable scenario may arise in SnO_2 by employing a more refined model. We begin by recalling the conditions required for the rigorous application of the DA. The first condition is that the potential must satisfy $qV_S \gg k_B T$, which defines a bulk region where the potential tends to zero. In the spherical case, however, this requirement alone is insufficient. It is possible for the depletion width to exceed the grain radius ($\Lambda > R$), in which case the potential cannot decay over the available radial length, but it would extend beyond it. Under such conditions, the bulk region disappears. The absence of a bulk region prevents defining the zero of potential at the grain center, which is consistent with the surface states formula that always refers to the Fermi level. The situation is described below.

If qV_S is not $\gg k_B T$, Poisson's equation becomes:

$$\frac{1}{r}\frac{d^2}{dr^2}[r\Phi(\mathrm{r})] = -\frac{qN_d}{\varepsilon} + \frac{qN_d}{\varepsilon}e^{\frac{q\Phi(\mathrm{r})}{k_B T}},$$

with the boundary conditions:

$$-\left.\frac{d\Phi(\mathrm{r})}{dr}\right|_{r=0} = 0, \quad \Phi(R) = -V_S.$$

Let us now consider a single grain and its internal double-barrier profile, with the potential zero defined in the bulk (as in Fig. 7.13). Referring to Fig. 7.16, imagine decreasing the grain radius until $R = \Lambda$. At this point, the bulk collapses to a single point at the center of the grain (where the potential is zero), and the two barriers meet at this point. If the grain radius is further reduced so that the new radius $R' < \Lambda$, the potential no longer has sufficient space to decay, and the central point where the two barriers meet shifts further away from zero as the radius decreases.

Let us consider the electron density:

$$n = N_d e^{\frac{q\Phi}{k_B T}},$$

where the reference for Φ is the conduction band energy in the bulk ($E_{C,\,\mathrm{bulk}} = 0$), such that $n = N_d$ in the bulk.At the center of the grain, however, $\Phi \neq 0$ for $R' < \Lambda$.

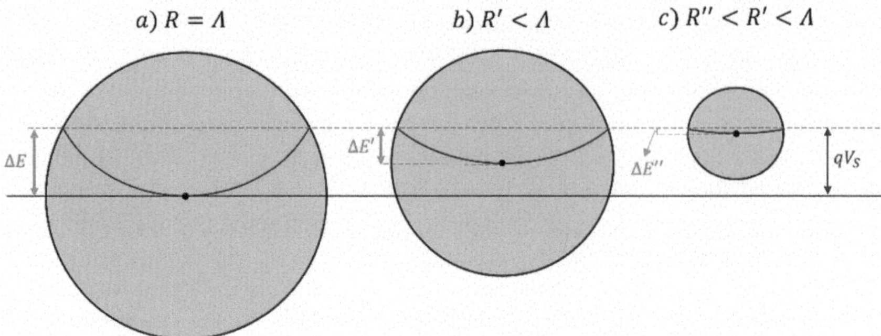

Fig. 7.16 **a** Grain with $R = \Lambda$, where the bulk collapses into a single point of zero energy; **b** grain with $R' < \Lambda$, where the potential has not space enough to extinguish at the grain center; **c** grain with $R'' < R'$, exhibiting barrier flattening

Thus, the potential difference between surface and center becomes very small:

$$\Delta\Phi = \Phi_{\text{surface}} - \Phi_{\text{center}} \ll V_S,$$

where V_S refers to the surface potential measured via the Arrhenius plot and referenced to the potential zero in the bulk.

As the grain radius decreases (transitioning from case (a) to (c) in Fig. 7.16), the slope of the potential barrier becomes increasingly shallow; for very small grains, it nearly flattens. Consequently, the electric field $F = d\Phi/dr$ is large for large grains and small for small grains. Thus, the field decreases with decreasing R, and the surface charge density ($\sigma_{\text{surface}} = F\varepsilon$) also decreases.

This phenomenon is known as **band-bending flattening**. In this regime there is no bulk, therefore the assumption $\Phi_{\text{bulk}} = 0$ is no longer valid, and the equations must be reformulated.

Now, the potential in the boundary conditions (as given in the previous system of equations) is not considered at the center, but rather on the surface (always equal to V_S, which is experimentally measurable). The field condition can be taken at $r = 0$, where the spherical grain symmetry dictates that the field vector has no preferred direction and must therefore be zero. The last condition is the neutrality condition (derived by Gauss' theorem): the charge density in a grain without a bulk must be integrated within the grain. Since the charge density depends on the potential, it is not straightforward to express it directly, but according to Gauss' law, it is equal to the flux of the electric field (multiplied by ε_0). The total charge inside the grain must equal (with opposite sign) the charge on the surface, ensuring neutrality:

$$\frac{1}{\varepsilon_0}\int_{V=\frac{4}{3}\pi R^3} \rho(r)dV = \frac{1}{\varepsilon_0}\int_{V=\frac{4}{3}\pi R^3} \left[\frac{qN_d}{\varepsilon} + \frac{qN_d}{\varepsilon}e^{\frac{q\Phi(r)}{k_BT}}\right]dV = \frac{1}{\varepsilon_0}\int_S qN_S dS.$$

By Gauss' law:

$$\frac{1}{\varepsilon_0}\int_{V=\frac{4}{3}\pi R^3}\left(\frac{qN_d}{\varepsilon}+\frac{qN_d}{\varepsilon}e^{\frac{q\Phi(r)}{k_BT}}\right)dV=\int_S\vec{E}\cdot\hat{n}dS.$$

Since:

$$\int_S\vec{E}\cdot\hat{n}dS=E\left(4\pi R^2\right)$$

and:

$$\frac{1}{\varepsilon_0}\int_S qN_S dS=\frac{qN_S}{\varepsilon_0}\left(4\pi R^2\right),$$

the electric field results:

$$E=\frac{qN_S}{\varepsilon_0}.$$

Since $E = -\, d\Phi/dR$, we obtain:

$$-\frac{d\Phi}{dR}=\frac{qN_S}{\varepsilon_0}$$

which provides the neutrality condition expressed in terms of the potential. This extended model therefore enables the description of grains whose dimensions are small compared with the depletion width.

7.4 DA Scanning Tunneling Spectroscopy

A **Scanning Tunneling Microscope (STM)** is a type of scanning probe instrument designed to image surfaces with atomic-scale resolution. It detects surface features using an ultra-sharp conductive tip capable of resolving structures smaller than 0.1 nm, with a depth resolution of approximately 0.01 nm (10 pm) as reported by Bai et al. [10]. The operating principle of STM is based on quantum tunneling. When the tip is positioned in extremely close proximity to the surface, the application of a bias voltage between the tip and the sample allows electrons to tunnel across the vacuum gap separating them. The tunneling current depends on the tip-sample separation, the applied voltage, and the local density of states of the sample. Data are collected by monitoring the tunneling current as the tip scans across the surface, and the resulting information is typically represented in the form of a topographic image. The STM system (Fig. 7.17a) operates under ultra-high vacuum conditions (base pressure $\approx 10^{-11}$ mbar) and can acquire images of samples under heating or in controlled gas atmospheres (up to $\approx 10^{-7}$ mbar). Scanning Electron Microscopy

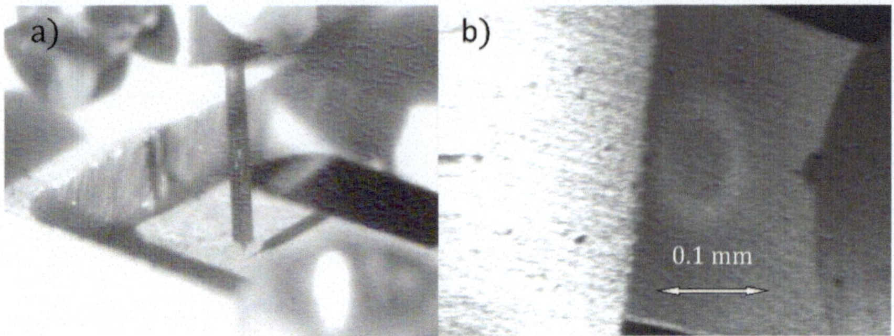

Fig. 7.17 **a** STM system magnification; **b** SEM employed to position the sample

(SEM) is employed to precisely position the tip over the sample surface, for example a SnO_2 film (0.1 mm × 0.1 mm), as shown in Fig. 7.17b.

Figure 7.18 presents a 3D STM image of a 200 × 200 nm^2 sample, acquired at room temperature with a bias of +3.2 V and a tunneling current of 0.6 nA. The color scale spans from 0 nm (black) to 28 nm (white), and the particle size is approximately 30 nm.

A variation of STM, known as **Scanning Tunneling Spectroscopy (STS)**, is performed by holding the tip fixed above a chosen location on the surface while sweeping the applied voltage and measuring the corresponding tunneling current. This technique enables the reconstruction of the local electronic density of states. In some cases, STS is performed under high magnetic fields or in the presence of impurities, allowing investigation of electron interactions and other quantum phenomena.

The examination of the sample's energy levels is carried out by adjusting the potential of the tip relative to the sample (ensuring that physical contact between tip

Fig. 7.18 3D STM image of a sample of 200 × 200 nm^2 with 30 nm of particle size, acquired at room temperature at +3.2 V and 0.6 nA; the color scale ranges from 0 nm (black) to 28 nm (white)

and sample never occurs). The sample remains electrically isolated and grounded, so its potential remains constant, while only the tip potential is modified by the operator. Adjusting the tip bias modifies the position of the tip Fermi level (E_F). For instance, applying a positive potential to the tip lowers its E_F; when the tip's Fermi level approaches the valence band top (VBT) of the semiconductor and slightly falls below the semiconductor's E_F, electrons from filled states in the semiconductor flow into empty states in the tip, generating an electronic current. Conversely, a negative tip bias raises the tip E_F; when it slightly exceeds the conduction band bottom (CBB), electrons begin to tunnel from the tip into the sample. All current transport occurs via quantum tunneling through the vacuum barrier due to the potential difference between the tip and the sample, no physical contact takes place. Figures 7.19 and 7.20 illustrate two situations: no applied tip bias (no tunneling current) and a positive tip bias (current flows from the sample to the tip).

For an ideal infinite crystal without surfaces (containing only Bloch states) no tunneling current flows when the tip Fermi level lies within the band gap, as no states exist in that energy range. In a real semiconductor, however, surface states appear

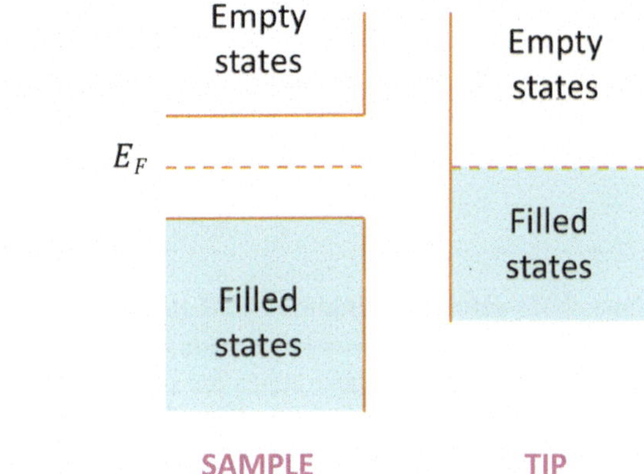

Fig. 7.19 Sample and tip states when no external potential is applied to the tip

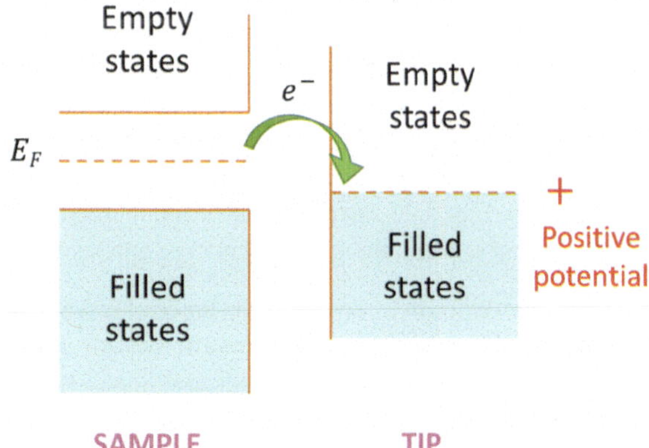

Fig. 7.20 Sample and tip states when a positive external potential is applied to the tip. Electron current flows from the sample to the tip

within the band gap, enabling a tunneling signal even when the tip's E_F is located inside the band gap. Specifically, if the tip's E_F is above surface acceptor states, electrons tunnel from the tip into the semiconductor; if it lies below surface donor states, electrons tunnel from the semiconductor into the tip. Thus, the presence of surface states yields a measurable tunneling current even within the band gap region. Sweeping the sample voltage from negative to positive therefore yields an IV curve containing local electronic information, including the positions of the band edges, gap states, and the local density of states structure, as reported by Feenstra [11].

Figure 7.21 shows the average of 6400 I/V curves. The upper curve corresponds to large grains (grain radius larger than the depletion width), whereas the lower curve corresponds to small grains (grain radius smaller than the depletion width). The surface barrier (S_B) for the 10 nm sample is approximately 0.1 eV lower than for the 30 nm sample. The y-axis is proportional to the surface state density (N_{SS}), while the x-axis represents the sample bias, i.e., the potential difference between the tip and the semiconductor. The "zero" point on the x-axis corresponds to the equilibrium condition with an unbiased tip. The distance between this zero and the onset of the conduction band, identified by a marked increase in current (dI/dV), gives the position of the conduction band at the surface. Why at the surface? Because the tip is not penetrating the material but is simply positioned at the surface. Therefore, this method enables precise detection of the surface values both of the valence and the conduction band edges (VBT and CBB). The surface value of the conduction band is given by:

$$E_C(\text{surface}) - E_F = S_B.$$

Thus, STS offers a straightforward means to determine both S_B and N_{SS}. From Fig. 7.21, it is evident that large grains exhibit a high density of states even within the band gap, whereas these states are strongly suppressed in small grains (the curve in the gap region becomes nearly flat). As expected, N_{SS} decreases as the grain radius decreases, and this reduction is accompanied by a decrease in S_B. This behavior is consistent with the unpinning effect. It is well established that L_0 remains constant when N_{SS} is large (i.e., when the Fermi level is pinned), such that $E_F - \Phi_0 = 0$ and $L_0 = S_B = \text{const}$, with $N_S = N_{SS}(E_F - \Phi_0)$. This situation corresponds to large grains, where experiments confirm that N_{SS} is high. In contrast, for small grains, N_{SS} is significantly reduced, allowing E_F to shift away from Φ_0. This explains why S_B is lower in small grains. When N_S is not sufficiently large and N_{SS} is finite, $E_F - \Phi_0 \neq 0$, and $S_B < L_0$. The observed decrease in S_B by approximately an order of magnitude in Fig. 7.21 can therefore be attributed to the unpinning of the Fermi level.

In summary, when the grain radius becomes small, two key effects emerge:

1. from the neutrality condition, this model predicts a decrease in N_S, leading to an enhanced sensing response;
2. experimentally, a significant decrease in N_{SS} has been observed, which results in Fermi level unpinning and an associated increase in material sensitivity.

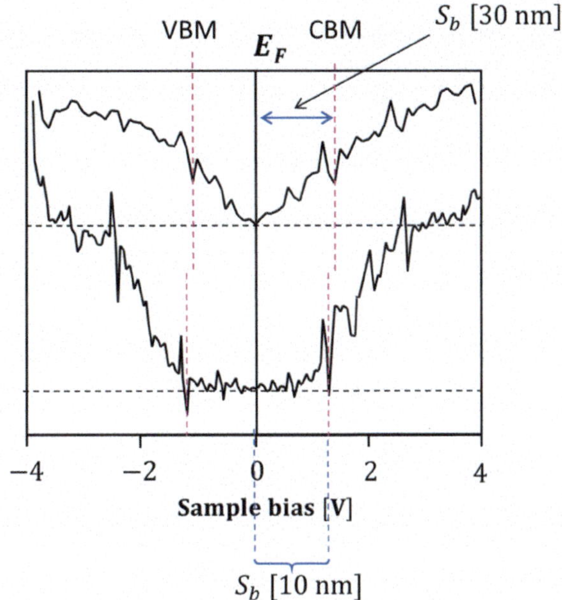

Fig. 7.21 Average of 6400 I/V curves. The sample bias is the potential difference between the tip and the semiconductor

7.5 Tunneling Approach

Up to this point, only thermionic emission across the barrier has been considered. At this stage, it is important to also account for the tunneling contribution to achieve a complete analysis. In a first approximation, the grains under consideration are assumed to be sufficiently large to contain a bulk region, i.e., the grain radius is much larger than the depletion width. Under this condition, the D.A. can be applied, and the planar Schottky relations are valid, since $\Lambda \ll R$. The probability of an electron tunneling through the barrier is evanescent. Figure 7.22 illustrates an electron with energy E, measured with respect to the CBB, impinging on the barrier at position x_1 and exiting at position x_2.

The well-known relation describing the double barrier (see the application of the D.A. for p-n junctions in Sect. 5.3) is given by:

$$qV(x) = \begin{cases} \frac{qV_S}{\Lambda^2}(x+\Lambda)^2, & x \leq 0, \\ \frac{qV_S}{\Lambda^2}(x-\Lambda)^2, & x \geq 0. \end{cases}$$

where $V_S = (qN_d/2\varepsilon)\Lambda^2$.

For electron energies $E < qV_S$, the wave vector k becomes imaginary, and it is convenient to define a real quantity β such that the wavefunction represents an evanescent wave:

$$\psi(x) \propto e^{ikx},$$

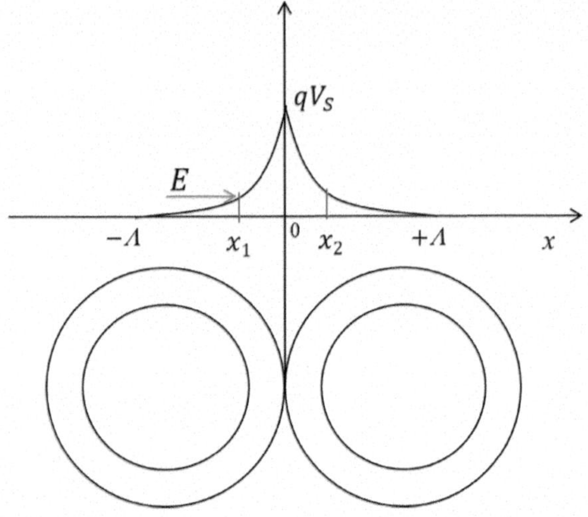

Fig. 7.22 An electron with energy E, with respect to CBB, impinges at x_1 and exits at x_2

with

$$k = \sqrt{\frac{2m^*}{\hbar^2}(E - qV)}.$$

Setting $k = i\beta$, one obtains:

$$\beta = \sqrt{\frac{2m^*}{\hbar^2}(qV - E)}$$

and the wave function becomes:

$$\psi(x) \propto e^{-\beta x}.$$

The tunneling probability T is therefore proportional to the squared modulus of the wavefunction:

$$|\psi(x)|^2 \propto e^{-2\beta x}.$$

Calculations are performed using the **WKB approximation** (from Wentzel-Kramers-Brillouin), a semiclassical method for solving linear differential equations with spatially varying coefficients. This method is widely employed in semiclassical quantum mechanics, where the wavefunction is expressed in exponential form and expanded within a semiclassical framework, under the assumption that either its amplitude or its phase varies slowly in space. The key assumption of the WKB approach is that the potential varies negligibly over a distance comparable to the wavelength of the impinging wave, allowing the semiclassical approximation to be applied. In this case, β is treated as a position-dependent quantity $\beta(x)$ and the potential is treated as a function of position x between the classical inversion points. This approach is regarded as semiclassical because, in quantum mechanics, momentum and position cannot be simultaneously defined with arbitrary precision and are therefore not compatible observables. In a fully quantum description, the electron energy is associated with the CBB and is expressed as a function of the wave vector $\vec{k}$. In the present treatment, however, the energy is effectively described as a function of the spatial coordinate x, an approximation that is permissible only within the semiclassical framework and is widely adopted in solid-state physics. Under these assumptions, the tunneling probability for an electron of energy E is:

$$T(E) = e^{-2\int_{x_1}^{x_2}\beta(x)dx} = e^{-2\int_{x_1}^{x_2}\sqrt{\frac{2m^*}{\hbar^2}(qV(x)-E)}dx}.$$

It can also be expressed as:

$$T(E) = \exp\left\{-2\frac{2m^*}{\hbar^2}\left[\int_{x_1}^{0}\sqrt{\frac{qV_S}{\Lambda^2}(x+\Lambda)^2 - Edx} + \int_{0}^{x_2}\sqrt{\frac{qV_S}{\Lambda^2}(x-\Lambda)^2 - Edx}\right]\right\}.$$

The barrier width is typically on the order of 20 – 40 nm, suggesting that the tunneling probability is small. However, electrons are distributed over a range of energies, and those with higher energies encounter a narrower effective barrier (since the barrier narrows with increasing energy). When all contributions are summed over energy, the tunneling current can become substantial, i.e., the tunneling contribution becomes important when it is integrated in energy. The two integrals in the expression above are equal by symmetry, although this might not be immediately obvious.

Let us express the integration limits as functions of Λ and E:

$$qV(x_1) = E = \frac{qV_S}{\Lambda^2}(x_1+\Lambda)^2,\ \ qV(x_2) = E = \frac{qV_S}{\Lambda^2}(x_2-\Lambda)^2,$$

yielding:

$$x_1 = -\Lambda\left[1-\left(\frac{E}{qV_S}\right)^{\frac{1}{2}}\right],\ \ x_2 = \Lambda\left[1-\left(\frac{E}{qV_S}\right)^{\frac{1}{2}}\right].$$

Thus, the two integrals to be solved are:

$$S_1 = \int_{x_1}^{0}\sqrt{\frac{qV_S}{\Lambda^2}(x+\Lambda)^2 - Edx},\ \ S_2 = \int_{0}^{x_2}\sqrt{\frac{qV_S}{\Lambda^2}(x-\Lambda)^2 - Edx},$$

They can be rewritten as:

$$S_1 = \int_{-\Lambda\left[1-\left(\frac{E}{qV_S}\right)^{\frac{1}{2}}\right]}^{0} \sqrt{(x+\Lambda)^2 - \frac{\Lambda^2 E}{qV_S}} dx \left(\frac{\sqrt{qV_S}}{\Lambda}\right),$$

and

$$S_2 = \int_{0}^{\Lambda\left[1-\left(\frac{E}{qV_S}\right)^{\frac{1}{2}}\right]} \sqrt{(x-\Lambda)^2 - \frac{\Lambda^2 E}{qV_S}} dx \left(\frac{\sqrt{qV_S}}{\Lambda}\right).$$

Then, performing a variable change in S_1:

$$u = x + \Lambda$$

$$du = dx.$$

If $x = -\Lambda[1 - (E/qV_S)^{1/2}]$, then $u = \Lambda(E/qV_S)^{1/2}$, and if $x = 0$, then $u = \Lambda$. Thus, S_1 becomes:

$$S_1 = \frac{\sqrt{qV_S}}{\Lambda} \int_{\Lambda\left(\frac{E}{qV_S}\right)^{\frac{1}{2}}}^{\Lambda} \sqrt{u^2 - \frac{\Lambda^2 E}{qV_S}} du.$$

Similarly, for S_2, performing the change of variable:

$$u = \Lambda - x$$

$$du = -dx.$$

If $x = \Lambda[1 - (E/qV_S)^{1/2}]$, then $u = \Lambda(E/qV_S)^{1/2}$ and if $x = 0$, then $u = \Lambda$ (as in the previous case). Thus, S_2 becomes:

$$S_2 = \frac{\sqrt{qV_S}}{\Lambda} \int_{\Lambda\left(\frac{E}{qV_S}\right)^{\frac{1}{2}}}^{\Lambda} \sqrt{u^2 - \frac{\Lambda^2 E}{qV_S}} du.$$

Therefore, one finds that $S_1 = S_2$, and the total integral is:

$$S_1 + S_2 = 2\frac{\sqrt{qV_S}}{\Lambda} \int\limits_{\Lambda\left(\frac{E}{qV_S}\right)^{\frac{1}{2}}}^{\Lambda} \sqrt{u^2 - \frac{\Lambda^2 E}{qV_S}} du.$$

Using the integral identity:

$$\int (x^2 - a^2)^{\frac{1}{2}} dx = \frac{x}{2}(x^2 - a^2)^{\frac{1}{2}} + \frac{a^2}{2} \ln\left[x + (x^2 - a^2)^{\frac{1}{2}}\right],$$

with $x^2 = u^2$ and $a^2 = \Lambda^2 E/qV_S$, we obtain:

$$S_1 + S_2 = \Lambda\sqrt{qV_S}\left\{\left(1 - \frac{E}{qV_S}\right)^{\frac{1}{2}} - \frac{E}{qV_S} \ln\left[\frac{1 + \left(1 - \frac{E}{qV_S}\right)^{\frac{1}{2}}}{\left(\frac{E}{qV_S}\right)^{\frac{1}{2}}}\right]\right\}.$$

This result shows a dependence on E/qV_S and the tunneling probability $T(E)$ for a single electron with energy E to cross the double parabolic barrier is:

$$T(E) = \exp\left\{-2\frac{2m^*}{\hbar^2}\Lambda\sqrt{qV_S}\left\{\left(1 - \frac{E}{qV_S}\right)^{\frac{1}{2}} - \frac{E}{qV_S} \ln\left[\frac{1 + \left(1 - \frac{E}{qV_S}\right)^{\frac{1}{2}}}{\left(\frac{E}{qV_S}\right)^{\frac{1}{2}}}\right]\right\}\right\}.$$

The factor of 2 at the beginning accounts for the presence of a double barrier (since the two integrals yield identical results). The effective mass m^* incorporates all the information about the material's band structure, emphasizing that electrons are not free particles but are confined within the crystal lattice.

Defining:

$$\alpha = \frac{E}{qV_S}$$

and

$$E_{00} = \frac{q\hbar}{2}\left(\frac{N_d}{m^*\varepsilon}\right)^{\frac{1}{2}},$$

and recalling that we are working with a planar geometry, so:

$$\Lambda = \sqrt{\frac{2\varepsilon V_S}{qN_d}},$$

by multiplying and dividing the integral by $\sqrt{q}$, we can compactly write the tunneling probability as:

$$T(E) = \exp\left[-2\frac{qV_S}{E_{00}}y(\alpha)\right],$$

where:

$$y(\alpha) = (1-\alpha)^{\frac{1}{2}} - \alpha \ln\left[\frac{1+(1-\alpha)^{\frac{1}{2}}}{\alpha^{\frac{1}{2}}}\right].$$

So far, we have derived the tunneling probability for an electron with a given energy E relative to the CBB. To account for all electrons, the tunneling probability must be integrated over the electron energy distribution. The flat-band current density is:

$$J_m = A^* T^2\, e^{\left[-\frac{(E_C-E_F)}{k_B T}\right]},$$

where A^* is Richardson constant. The total current density is given by the integral of the tunneling probability multiplied by the electron distribution at a specific energy:

$$J_f = \frac{A^* T}{k_B}\int_0^\infty T(E) e^{-\frac{E+E_C-E_F}{k_B T}}\, dE,$$

where the integration runs from the CBB ($E = 0$) to the vacuum energy ($E \to \infty$), and the exponential term represents the electron distribution, indicating how many electrons possess a specific energy E relative to the CBB.

Normalizing with respect to J_m, one obtains:

$$\frac{J_f}{J_m} = \frac{1}{k_B T}\int_0^\infty T(E) e^{-\frac{E}{k_B T}}\, dE,$$

a dimensionless quantity. This normalization removes the explicit dependence on the Fermi level, while the exponential factor reflects the energy distribution of electrons with energy measured relative to the CBB.Splitting the integral yields:

$$\frac{J_f}{J_m} = \frac{1}{k_B T}\int_0^{qV_S} T(E) e^{-\frac{E}{k_B T}}\, dE + \frac{1}{k_B T}\int_{qV_S}^\infty e^{-\frac{E}{k_B T}}\, dE,$$

where:

$$I_1 = \frac{1}{k_B T}\int_0^{qV_S} T(E)e^{-\frac{E}{k_B T}}dE, \quad I_2 = \frac{1}{k_B T}\int_{qV_S}^{\infty} e^{-\frac{E}{k_B T}}dE,$$

so

$$\frac{J_f}{J_m} = I_1 + I_2.$$

The first term, I_1, represents the **pure tunneling contribution**, corresponding to the current carried by electrons with energies lower than the barrier height. The second term, I_2, represents the **thermionic contribution** and involves electrons with energies higher than the barrier. This formulation therefore provides a direct measure of the relative importance of tunneling and thermionic emission.

The second term is straightforward to evaluate. In this case, $T(E) = 1$ for $E > qV_S$, since no barrier is present beyond this energy, yielding the conventional thermionic contribution (i.e., the Boltzmann factor normalized to the flat-band case):

$$I_2 = \frac{1}{k_B T}\int_{qV_S}^{\infty} e^{-\frac{E}{k_B T}}dE = e^{-\frac{E}{k_B T}}.$$

For the tunneling contribution, we have:

$$I_1 = \frac{1}{k_B T}\int_0^{qV_S} T(E)e^{-\frac{E}{k_B T}}dE = \frac{1}{k_B T}\int_0^{qV_S} e^{-2\frac{qV_S}{E_{00}}y(\alpha)}e^{-\frac{E}{k_B T}}dE,$$

where $\alpha = \frac{E}{qV_S}$ and $dE = qV_S d\alpha$. The integral can be rewritten as:

$$I_1 = \frac{qV_S}{k_B T}\int_0^1 e^{\left(-2\frac{qV_S}{E_{00}}y(\alpha) - \frac{\alpha qV_S}{k_B T}\right)}d\alpha = \frac{qV_S}{k_B T}\int_0^1 e^{-\frac{qV_S}{k_B T}\left[\frac{2k_B T y(\alpha)}{E_{00}} + \alpha\right]}d\alpha.$$

By comparing the values of $k_B T$ and E_{00}, it is possible to identify which of the two contributions is dominant (both being dimensionless owing to the normalization). Using experimentally values, it is found that, at the sensor operating temperature, the tunneling contribution is of the same order of magnitude as the thermionic contribution.

In Fig. 7.23, an example of three pairs of curves is reported, where the y-axis represents $(dI/dE)/(dI/dE)_{max}$, i.e., the current variation as a function of energy. Each pair consists of a pink curve, corresponding to $E_B/(k_B T) = 40$, and a green curve, corresponding to $E_B/(k_B T) = 10$, while the ratio $(k_B T)/E_{00}$ increases along the x-axis (E/E_B). The pink curves are noticeably narrower than the green ones.

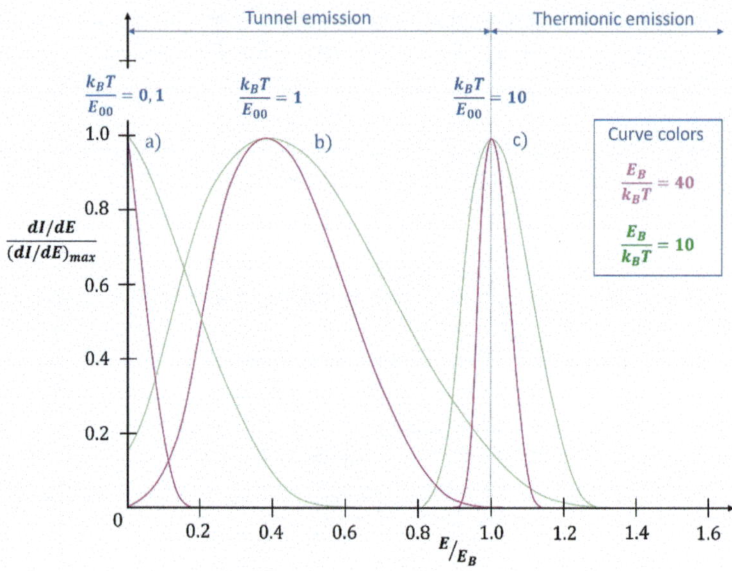

Fig. 7.23 Example of couple of curves, each one composed by a pink ($E_B/(k_BT) = 40$) and a green curve ($E_B/(k_BT) = 10$) where the ratio k_BT/E_{00} increases along the x axis **a** 0.1; **b** 1; **c** 10

When $E/E_B > 1, 0$, meaning that the electron energy E exceeds the barrier energy E_B (which is fixed in height), the current contribution becomes purely thermionic. Considering the curves in group (b), where $(k_BT)/E_{00} = 1$ (ambient temperatures), it is evident that, due to the asymmetry of the curves (more elongated on the right side), at higher temperatures (green curve), both tunneling and thermionic emission are present, with tunneling dominating. Focusing instead on group (c), where $(k_BT)/E_{00} = 10$ (typical sensor operating temperatures), at higher E/E_B these curves are narrower than curves in group (b), so the asymmetry is less evident, and the contribution to the current is approximately halfway between tunnel and thermionic (the second slightly dominant).

As the sensor temperature decreases, the thermionic contribution is progressively reduced and the curves become narrower, while the barrier height remains constant.

In Fig. 7.24, the barrier profiles corresponding to two different grain configurations are shown. In case (a), the grain contains a bulk region, since the grain radius is larger than the depletion width ($R < \Lambda$). In case (b), the grain is fully depleted, as the radius equals the depletion width ($R = \Lambda$). Although the Schottky barrier height is the same in both configurations, the barrier shape differs as a consequence of the variation in $\Delta E = E_C - E_0$, where E_C denotes the conduction band energy at the surface and E_0 the conduction band energy at the center of the grain.

The thermionic contribution to the current is governed by the electron density at the surface, which can be expressed as:

$$n_S = N_C e^{\frac{E_F - E_0 + E_0 - E_C}{k_BT}} = n_0 e^{-\frac{\Delta E}{k_BT}},$$

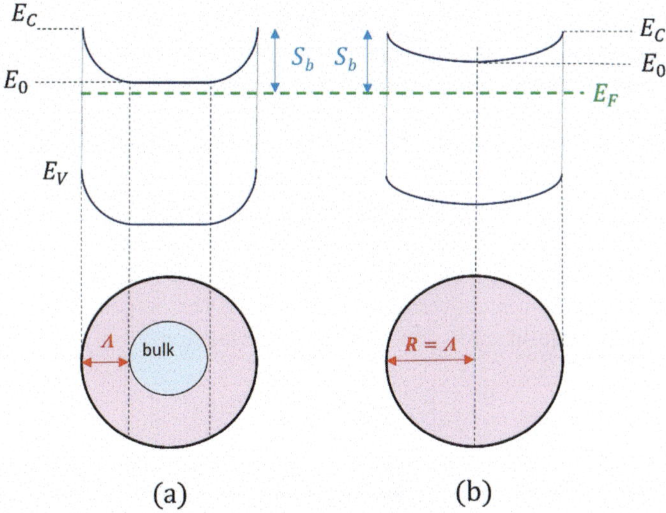

Fig. 7.24 Barrier shape of **a** grain with a bulk ($R < \Lambda$); **b** grain without a bulk ($R = \Lambda$), so fully depleted. The Schottky barrier remains the same but the shape changes, because it changes $\Delta E = E_C - E_0$, where E_C is the CB energy at the surface and E_0 is the CB energy at the grain center

where n_0 is the electron density at the center of the grain:

$$n_0 = N_C e^{\frac{E_F - E_0}{k_B T}}.$$

The thermionic contribution is independent of the grain radius R. In case (a), $\Delta E = qV_S$ and $n_0 = N_d$, whereas in case (b), $\Delta E < qV_S$ and $n_0 \ll N_d$, such that n_S remains unchanged provided that S_B is constant. Consequently, the thermionic contribution depends exclusively on the Schottky barrier height and not on the band bending or the grain radius. For small grains, $\Delta E \approx 0.2 - 0.3$ eV, whereas for larger grains $\Delta E \approx 1$ eV, indicating a substantial variation.

It must be demonstrated that the thermionic contribution does not depend on the shape of the barrier but only on S_b, whereas the tunneling contribution is shape dependent. Tunneling current is favored when the barrier is small, as this reduces the distance the electrons must traverse compared to larger barriers.

7.6 In-diffusion

It has been experimentally observed in metal oxides that variations in the electrostatic potential, which directly affect the barrier height, are reflected in measurable changes in the sensor response. A fundamental phenomenon underlying this behavior is the continuous exchange of oxygen between the surface and the bulk of the material. This exchange is intimately linked to the formation and annihilation of oxygen vacancies and, consequently, to the effective doping level of the semiconductor.

Specifically, oxygen molecules from the surrounding atmosphere can be adsorbed on the grain surface (O_{ads}) and subsequently diffuse into the material, where they become interstitial oxygen (O_{int}). Once inside the lattice, interstitial oxygen may recombine with a doubly ionized oxygen vacancy (V^{++}), thereby neutralizing the vacancy and returning the two electrons previously donated by the vacancy to the conduction band. This mechanism is referred to as **oxygen vacancy annihilation**. Because the interaction between the gas phase and the solid is continuous, vacancy creation and annihilation occur simultaneously. At a given temperature T, the system reaches a dynamic equilibrium, which follows the equation:

$$V^{++} + 2e^{-} + O_{\text{int}} \rightleftarrows O_{\text{lattice}}.$$

This process, schematically illustrated for SnO_2 in Fig. 7.25, leads to a modification of the doping level N_d of the material. It can be described as a sequence of three coupled steps:

(a) the establishment of an oxygen exchange equilibrium between the atmosphere and the grain surface;
(b) the diffusion of oxygen into and out of the grain;
(c) the creation and recombination of oxygen interstitial vacancies within the lattice.

Intuitively, an increase in the oxygen partial pressure in the atmosphere shifts the reaction towards oxygen adsorption, which in turn promotes vacancy annihilation. The vacancies annihilation leads to a transformation in the semiconductor's band structure.

Within the present model, in-diffusion is assumed to modify the bulk doping without significantly altering the surface electronic structure that fixes the S_b; this assumption is supported by experimental evidence in metal oxide systems but may fail if surface chemistry is substantially modified as will be shown in Sect. 7.7.

If the barrier height S_b is not affected by in-diffusion, the thermionic contribution (I_2) to the total current remains unchanged. However, the tunneling contribution is reduced due to the altered curvature of the barrier, which becomes relatively larger than before for electrons at the same distance from the Fermi level. This explains the gradual increase in resistance after the surface equilibrium is reached.

To clarify this behavior, it is useful to recall the neutrality condition $N_S = \Lambda N_d$. If N_S is kept constant due to Fermi-level pinning, a decrease in N_d necessarily leads to an increase in Λ. A similar conclusion can be reached even in the absence of pinning by using the Schottky relation. An increase in oxygen partial pressure corresponds to an increase in Λ; more oxygen enters the material, "killing" vacancies. Consequently, a wider depletion region is required to maintain neutrality. More charge is needed inside the grain to neutralize the surface charge. As a result, by increasing the oxygen partial pressure, even large grains become fully depleted, leading to a modification in the band bending, as observed earlier. In both scenarios (a) and (b) in Fig. 7.26, the thermionic current contribution remains unchanged because it depends solely on S_b

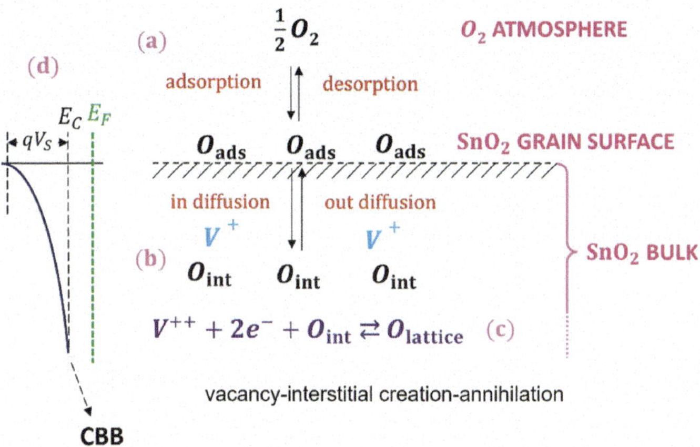

Fig. 7.25 In- and out-diffusion mechanisms in an *n*-type metal oxide (e.g., SnO_2). **a** Oxygen exchange equilibrium between the atmosphere and the grain surface; **b** oxygen diffusion into and out of the grain; **c** interstitial vacancy creation and recombination; **d** corresponding energy band diagram

and not on the band bending. In contrast, the tunneling contribution is significantly affected by this modification, as it is reduced due to the relative enlargement of the barrier. In-diffusion only alters the doping level N_d (at fixed N_S and N_{SS}, so that $E_F - \phi_0$ is constant). N_{SS} remains constant because N_d does not influence the surface band structure. This provides a possible explanation for the significant change in conductivity when oxygen levels vary. It also accounts for the observed gradual increase in resistance after surface equilibrium is reached, as seen in Fig. 7.27 for SnO_2, at a working temperature of 400 °C and exposed to synthetic dry air (20% O_2 and 80% N_2).

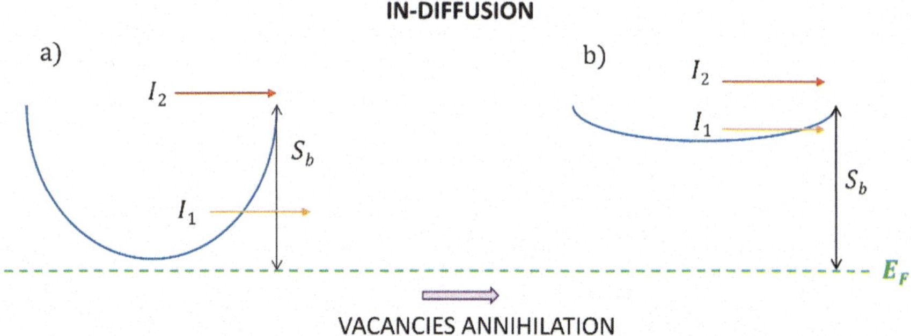

Fig. 7.26 Vacancy annihilation due to oxygen in-diffusion. The Schottky barrier height remains unchanged; therefore, the thermionic contribution is unaffected by changes in barrier curvature. The tunneling contribution, however, is larger in case (**a**), where the barrier is thinner, and reduced in case (**b**), where the barrier becomes wider

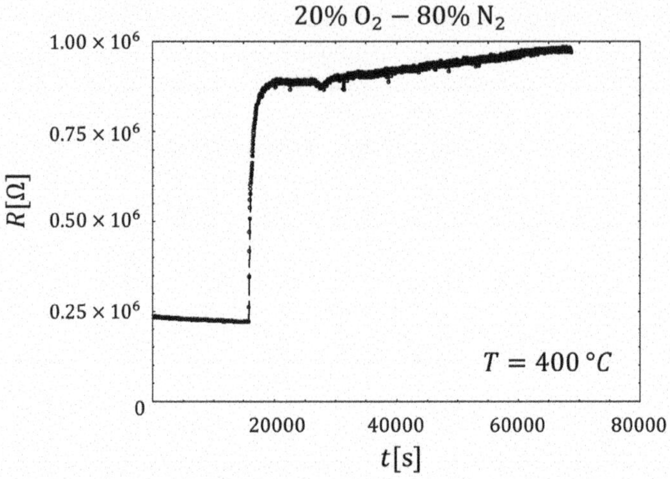

Fig. 7.27 Resistance variation (measured in Ω) of SnO_2 set at a working temperature $T = 400$ °C and exposed to synthetic dry air (20% O_2 and 80% N_2). The response shows a slow drift during time (slow increase), attributed to the in-diffusion phenomenon

7.7 Oxygen Vacancies Dependence on the Fermi Level and Non-parabolic Barriers

As illustrated in Fig. 7.24, the processes of vacancy creation and annihilation arising from oxygen in- and out-diffusion are governed by the following reaction:

$$O_0 \rightleftarrows V^{++} + 2e^- + \frac{1}{2}O_2,$$

where V^{++} denotes a doubly ionized vacancy, O_2 represents an interstitial oxygen (located inside the crystal but spatially separated from the vacancy), and O_0 is an oxygen atom belonging to the lattice. The relationship between the concentrations of the involved species is expressed as:

$$K = \left[V^{++}\right]n^2 p(O_2)^{\frac{1}{2}},$$

where K is the concentration of lattice oxygen, n is the electron density, and $p(O_2)$ is the oxygen partial pressure. Let us now derive this equation step by step. In the first step, an oxygen atom (O_S^O) is removed from a lattice site within the crystal and transferred to the surface, where S denotes a surface site:

$$O_0 + S \rightleftarrows V + O_S^0.$$

The corresponding relationship between the concentrations is:

$$\frac{[V][O_S^0]}{[S]} \propto e^{\frac{-E_{\text{surf}}}{k_B T}},$$

where E_{surf} is the energy required for this process. The second step involves oxygen adsorption at the surface:

$$O_S^0 \rightleftarrows \frac{1}{2} O_2 + S,$$

for which the concentration relationship is:

$$\frac{p(O_2)^{\frac{1}{2}}[S]}{[O_S^0]} \propto e^{\frac{-E_{\text{ads}}}{k_B T}},$$

where E_{ads} is the energy required for this adsorption process. By combining these two equations, the vacancy concentration can be written as:

$$[V] \propto p(O_2)^{-\frac{1}{2}} e^{\frac{-(E_{\text{sur}}+E_{\text{ads}})}{k_B T}}.$$

Defining the formation energy $E_{\text{for}} = E_{\text{sur}} + E_{\text{ads}}$, which represents the total energy required to transfer an oxygen atom from a lattice site inside the crystal to the gas phase, the expression becomes:

$$[V] \propto p(O_2)^{-\frac{1}{2}} e^{\frac{-E_{\text{for}}}{k_B T}}.$$

This result shows that the vacancy concentration decreases with increasing oxygen partial pressure and increasing formation energy.

In Fig. 7.28, the dependence of the defect formation enthalpy ΔH_d (panel c) on the Fermi energy E_F is shown. In panel (a), an occupied donor D^0 with energy $E_D < E_F$ is electrically neutral; consequently, ΔH_d does not depend on E_F and is represented by a horizontal line in panel (c). In panel (b), the donor is unoccupied ($E_D > E_F$) and therefore positively ionized; the electron released from the donor is transferred to the Fermi level, resulting in an energy gain δ. For this reason, the defect formation enthalpy increases as the Fermi level decreases, and the corresponding line in panel (c) acquires a finite slope. In panel (c), the dependence of ΔH_d on E_F is reported for both donor and acceptor states. The crossing between the neutral and charged defect states identifies the energy position of the defect levels E_D [12].

To determine the concentration of doubly ionized vacancies $[V^{++}]$, the two electrons (denoted by indices 1 and 2) released to the Fermi level result in an energy gain:

$$\delta_1 + \delta_2 = (E_{D1} - E_F) + (E_{D2} - E_F) = E_{D1} + E_{D2} - 2E_F$$

where E_{D1} and E_{D2} are the two donor energy levels. By defining:

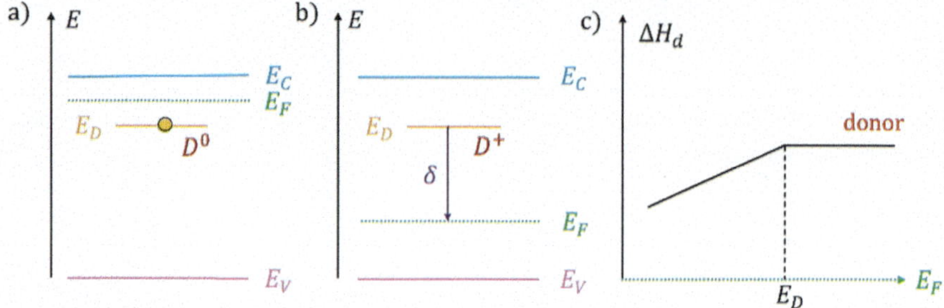

Fig. 7.28 **a** Occupied donor D^0 of energy $E_D < E_F$; **b** unoccupied donor of energy $E_D > E_F$, electron is released to the reservoir E_F leading to an energy gain δ; **c** dependence of the defect formation enthalpy ΔH_d (for a donor state) on the Fermi Energy E_F

$$E_0 = E_{\text{for}} + 2E_C - E_{D1} - E_{D2},$$

where E_0 represents the intrinsic energetic cost of forming a doubly ionized vacancy, excluding the explicit dependence on the Fermi level E_F, and solving for E_{for}:

$$E_{\text{for}} = E_0 - 2E_C + E_{D1} + E_{D2}$$

the concentration of doubly ionized vacancies is given by:

$$[V^{++}] \propto p(O_2)^{-\frac{1}{2}} e^{\frac{-[E_0 - 2(E_C - E_F)]}{k_B T}}.$$

This expression clearly indicates that, at fixed oxygen partial pressure and under equilibrium conditions, the density of ionized oxygen vacancies increases exponentially as E_F moves downward within the band gap, making it **impossible to achieve a uniform doping level** across the entire space-charge region. Considering now the bulk vacancy density N_d:

$$N_d \propto p(O_2)^{-\frac{1}{2}} e^{\frac{-[E_0 - 2(E_{CB} - E_F)]}{k_B T}}.$$

where E_{CB} is the conduction-band energy in the bulk, we obtain:

$$[V^{++}] \propto N_d e^{\frac{[2(E_C(x) - E_{CB})]}{k_B T}}.$$

This result, obtained by Romppainen and Lantto, 1988 through a different approach (drift-diffusion), can be simplified further [13]. Since $E_C(x) - E_{CB} = eV(x)$, it follows that:

$$\left[V^{++}\right] \propto N_d e^{\frac{2eV(x)}{k_B T}}.$$

Next, Poisson's equation becomes:

$$\frac{d^2V(x)}{dx^2} = \frac{2e}{\varepsilon}N_d e^{\frac{2eV(x)}{k_BT}}$$

which leads to the potential distribution:

$$V(x) = \frac{k_BT}{2\varepsilon}\ln\left[1+\tan^2\left(\frac{x}{\sqrt{2}L_D}\right)\right]$$

where the Debye length is given by:

$$L_D = \sqrt{\frac{\varepsilon k_BT}{4e^2N_d}}.$$

In Fig. 7.28, the potential profile near the surface due to band bending is shown, where x is the spatial coordinate as depicted in Fig. 7.29. The normalized coordinate x', expressed as a function of L_D, is given by:

$$x' = \frac{x}{\sqrt{2}L_D}.$$

The depletion-region width is obtained from charge neutrality as:

$$x_0 = \sqrt{2}L_D\tan^{-1}\left(\frac{N_S}{2\sqrt{2}N_dL_D}\right).$$

Accordingly, the height of the surface potential barrier is (Fig. 7.30):

$$V_S = \frac{k_BT}{2e}\ln\left[1+\frac{N_S^2}{8N_d^2L_D^2}\right].$$

Moreover, an **equation** can be derived **to determine the doping level and the position of the Fermi level**. From equilibrium, we know that:

$$N_d = Cp(O_2)^{-\frac{1}{2}}e^{-\frac{E_0-2(E_{CB}-E_F)}{k_BT}}$$

where C is a constant. For a general solid, with N_C being the effective density of states in the conduction band, the conduction electron distribution is given by:

$$n = N_C e^{-\frac{E_C-E_F}{k_BT}}.$$

where N_C is the effective density of states in the conduction band.For simplicity, we assume that at the operating temperatures considered the dominant donor centers are effectively ionized; however, for deeper levels or lower temperatures, partial ionization may occur and would require an explicit treatment. Assuming that, $n =$

$2N_d$, one obtains:

$$E_{CB} - E_F = \frac{E_0 - k_B T \ln\left[\frac{1Cp(O_2)^{-1/2}}{N_C}\right]}{3}.$$

To derive the **expression for the density of chemisorbed oxygen**, monoatomic oxygen adsorption is assumed at working temperatures $T > 150\,°\mathrm{C}$. Recalling:

$$O_S^0 \rightleftarrows \frac{1}{2}O_2 + S$$

and assuming low surface coverage, the surface site concentration $[S]$ can be treated as a constant. Thus, the earlier formula connecting the concentrations in the surface adsorption process simplifies to:

$$\frac{p(O_2)^1/2}{[O_S^0]} \propto e^{\frac{-E_{ads}}{k_B T}}.$$

The ratio of ionized to neutral oxygen atoms at the surface depends on the position of the Fermi level as:

$$\frac{[O_S^-]}{[O_S^0]} \propto e^{-\frac{E_t - E_F}{k_B T}},$$

where E_t is the **ionization energy**, which represents the energy of an electron at the top of the surface barrier. Therefore, the surface oxygen concentration N_S is:

$$N_S = Ap(O_2)^{\frac{1}{2}} e^{-\frac{eV_S + E_{CB} - E_F - E_{ads}}{k_B T}}.$$

To calculate the Schottky barrier height (Fig. 7.31), the values reported by Meier and Gopel (1988) were $E_{ads} = 2$ eV and $E_0 = 2.2$ eV [14].

In deriving these equations, it was assumed that, for a given oxygen pressure p_0 at $T = 600$ K, $N_d = 10^3\ \mathrm{m}^{-3}$ and $N_S = 10^{17}\ \mathrm{m}^{-3}$. These values were selected because they yield barrier heights commonly observed in practice. Finally, the Schottky barrier height is determined by the sum $(E_{CB} - E_F) + eV_S$ [15].

Assuming that all oxygen vacancies are doubly ionized, the bulk Fermi level position within the band gap can be determined from the electron concentration, expressed as $n = N_c \exp\left[-(E_C - E_F)/k_B T\right]$. Once the energy separation $E_{CB} - E_F$ is known, the constants C and A, appearing respectively in the expressions for the donor concentration N_d and the surface state density N_S, can be evaluated with the aid of the surface potential equation V_S.

Other results, reported by Barsan et al., 2011, show a change in work function of approximately 53 meV at 300 °C after a following a relative oxygen-pressure variation by a factor of 20 [16].

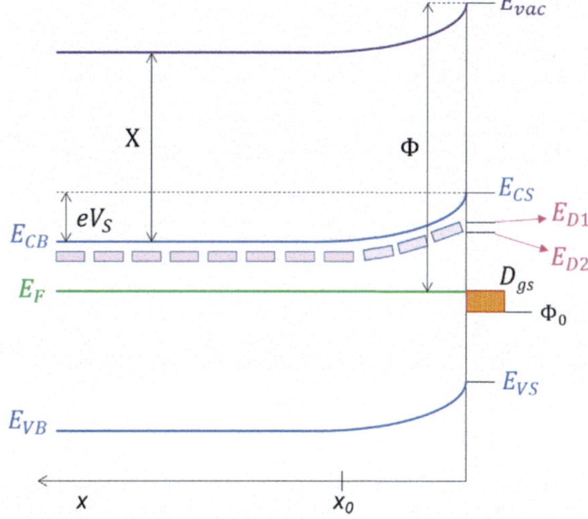

Fig. 7.29 Energetic configuration with two defect levels

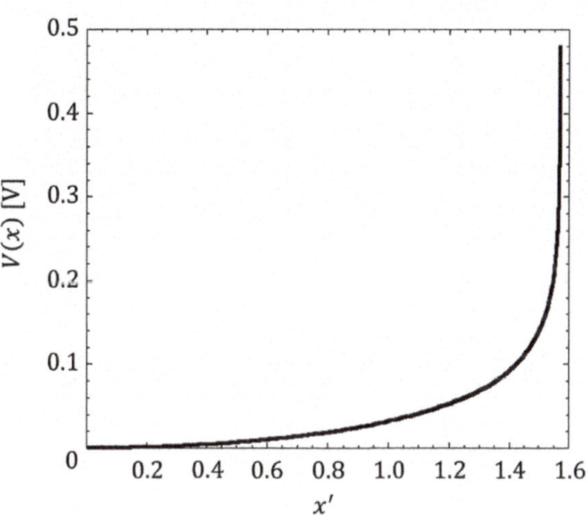

Fig. 7.30 Profile of the potential $V(x)$, expressed in V near the surface due to band bending at a working temperature $T = 600$ K

Changes in work function can be considered to directly reflect changes in the Schottky barrier height. In our model, a similar change in relative pressure leads to a change in the Schottky barrier height of 58 meV, a result that appears noteworthy given the assumptions, approximations, and experimental complexities involved.

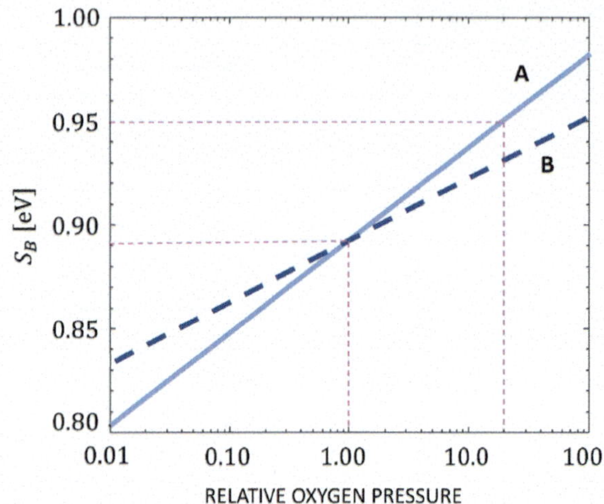

Fig. 7.31 Schottky barrier height (S_B) as a function of the relative oxygen pressure before (curve B) and after (curve A) diffusion

References

1. Samson S, Fonstad CG (1973) Defect structure and electronic donor levels in stannic oxide crystals. J Appl Phys 44:4618–4621
2. Munnix S, Schmeits M (1985) Origin of defect states on the surface of TiO_2. Phys Rev B 31:3369–3376
3. Chang SC (1980) Oxygen chemisorption on tin oxide: correlation between electrical conductivity and EPR measurements. J Vac Sci Technol 17:366–370
4. Lantto V, Romppainen P, Leppävuori S (1988) A study of the temperature dependence of the barrier energy in porous tin dioxide. Sens Actuators 14:149–158
5. Clifford PK, Tuma DT (1982/83) Characteristics of semiconductor gas sensors I. Steady state gas response. Sens Actuators 3:255–260
6. Carotta MC, Dallara C, Martinelli G, Passari L (1991) CH_4 thick-film gas sensors: characterization and modeling. Sens Actuators B Chem 3:191–196
7. Malagù C, Guidi V, Stefancich M, Carotta MC, Martinelli G (2002) A model for Schottky barrier and surface states in nanostructured n-type semiconductors. J Appl Phys 91:808–814
8. Martinelli G, Carotta MC (1995) Thick film gas sensors. Sens Actuators B Chem 23:157–161
9. Tang H, Prasad K, Sanjinès R, Schmid PE, Lévy F (1994) Electrical and optical properties of TiO_2 anatase thin films. J Appl Phys 75:2042–2047
10. Bai C (2000) Scanning tunneling microscopy and its applications. Springer, New York. ISBN 978-3-540-65715-6
11. Feenstra RM (1989) Surface electronic structure of semiconductor surfaces studied by scanning tunneling microscopy. Phys Rev Lett 63:1412–1415
12. Oba F et al (2010) Defect energetics in TiO_2 from first principles. J Phys Condens Matter 22:384211
13. Romppainen P, Lantto V (1988) Electrical transport mechanisms in porous semiconductor films. J Appl Phys 63:5159–5165
14. Maier J, Gopel W (1988) Ionic and electronic transport in oxides. J Solid State Chem 72:293–299
15. Aldao CM, Malagù C (2012) Non-parabolic intergranular barriers in tin oxide and gas sensing. J Appl Phys 112:024518
16. Barsan N, Hübner M, Weimar U (2011) Conduction mechanisms in metal oxide gas sensors. Sens Actuators B Chem 157:510–517

17. Malagù C, Carotta MC, Guidi V, Martinelli G (2004) Unpinning of Fermi level in nanocrystalline semiconductors. Appl Phys Lett 84:4158–4160
18. Malagù C, Carotta MC, Comini E, Faglia G, Giberti A, Guidi V, Maffeis TGG, Martinelli G, Sberveglieri G, Wilks SP (2005) Photo-induced unpinning of Fermi level in WO3. Sensors 5:594–603
19. Malagù C, Martinelli G, Ponce MA, Aldao CM (2008) Unpinning of the Fermi level and tunneling in metal oxide semiconductors. Appl Phys Lett 92:162104
20. Maffeis TGG, Owen GT, Malagù C, Martinelli G, Kennedy MK, Kruis FE, Wilks SP (2004) Direct evidence of the dependence of surface state density on the size of SnO2 nanoparticles observed by scanning tunnelling spectroscopy. Surf Sci 550:21–25
21. Malagù C, Carotta MC, Galliera S, Guidi V, Maffeis TGG, Martinelli G, Owen GT, Wilks SP (2004) Evidence of bandbending flattening in 10 nm polycrystalline SnO_2. Sens Actuators B 103:50–54
22. Malagù C, Carotta MC, Morandi S, Gherardi S, Ghiotti G, Giberti A, Martinelli G (2006) Surface barrier modulation and diffuse reflectance spectroscopy of MoO3-WO3 thick films. Sens Actuators B 118:94–97
23. Malagù C, Giberti A, Morandi S, Aldao CM (2011) Electrical and spectroscopic analysis in nanostructured SnO_2: "Long-term" resistance drift is due to in-diffusion. J Appl Phys 110:093711
24. Ponce MA, Malagù C, Carotta MC, Martinelli G, Aldao CM (2008) Gas in-diffusion contribution to impedance in tin oxide thick films. J Appl Phys 104:054907
25. Giberti A, Malagù C (2013) Current-voltage characteristics of nanostructured SnO2 films. Thin Solid Films 584:683–688
26. Morandi S, Prinetto F, Di Martino M, Ghiotti G, Lorret O, Tichit D, Malagù C, Vendemiati B, Carotta MC (2006) Synthesis and characterisation of gas sensor materials obtained from Pt/Zn/Al layered double hydroxides. Sens Actuators B 118:215–220
27. Aldao CM, Mirabella DA, Ponce MA, Giberti A, Malagù C (2011) Role of intragrain oxygen diffusion in polycrystalline tin oxide conductivity. J Appl Phys 109:063723
28. Giberti A, Carotta MC, Malagù C, Aldao CM, Castro MS, Ponce MA, Parra R (2011) Permittivity measurements in nanostructured TiO2 gas sensors. Phys Status Solidi A 208:118–122
29. Krik S, Gaiardo A, Valt M, Fabbri B, Malagù C, Guidi V (2021) First-principles study of electronic conductivity, structural and electronic properties of oxygen-vacancy-defected SnO_2. J Nanosci Nanotechnol 21
30. Krik S, Valt M, Gaiardo A, Fabbri B, Spagnoli E, Caporali M, Malagù C, Bellutti P, Guidi V (2022) Elucidating the ambient stability and gas sensing mechanism of nickel-decorated phosphorene for NO2 detection: a first-principles study. ACS Omega 7(11):9808–9817
31. Castro-Hurtado I, Malagù C, Morandi S, Pérez N, Mandayo GG, Castaño E (2013) Properties of NiO sputtered thin films and modeling of their sensing mechanism under formaldehyde atmospheres. Acta Mater 61:1146–1153
32. Wagner T, Kohl CD, Morandi S, Malagù C, Donato N, Latino M, Neri G, Tiemann M (2012) Photoreduction of mesoporous In2O3: Mechanistic model and utility in gas sensing. Chem Eur J 18:8216–8223
33. Shur M (1990) Transport phenomena in semiconductor devices. In: Physics of semiconductor devices. Prentice Hall, New York, pp 140–199

Chapter 8
Sensors Fabrication

> Design is not just what it looks like and feels like. Design is how it works.
> (Steve Jobs)

Semiconductor chemoresistive sensors represent one of the most established and widely employed classes of gas-sensing devices. They are typically based on ceramic materials composed of nanostructured semiconducting metal oxides (MOX) or metal sulfides, which exhibit good electrical stability at elevated operating temperatures. Their sensing mechanism relies on changes in the electrical resistance of the sensing film induced by surface chemical reactions with gaseous analytes. For this reason, nanostructuring plays a crucial role, as it maximizes the surface-to-volume ratio and, consequently, enhances the sensitivity of the device.

To be considered high-performing, a gas sensor must satisfy several key requirements:

- high sensitivity, i.e., a low detection limit for the target analyte;
- high selectivity, enabling discrimination against interfering gases;
- repeatability and long-term stability of the response;
- reproducibility of the fabrication process and of the response;
- low cost and low power consumption;
- ease of use;
- durability and operational robustness;
- suitability for miniaturization;
- compatibility with electronic systems for signal acquisition and processing.

The design and development of sensors capable of fulfilling these requirements constitute the main objective of current research in this field. Chapter 8 will focus on the chemoresistive gas sensors developed at the Sensor Laboratory (SL) of the Department of Physics and Earth Sciences of the University of Ferrara, describing their synthesis routes, device assembly, and functional characteristics. This Chapter draws upon an integrated analysis of several sources cited here [1–90], complemented by the authors' original contributions, in particular in sensor production and

© The Author(s), under exclusive license to Springer Nature Switzerland AG 2026

C. Malagù and G. Zonta, *Gas Sensors*, https://doi.org/10.1007/978-3-032-21614-4_8

experimental descriptions carried out at SL, with some specific references discussed at appropriate points in the text regarding material types.

8.1 Thick-Film Sensors

Chemoresistive gas sensors can be broadly divided into **thick film** and **thin film** devices according to the thickness of the sensing layer. This classification, summarized in Table 8.1, has significant implications for the resulting morphology, fabrication techniques, and overall sensing performance.

- **Thick films** ($\approx 1 - 50\ \mu m$) are generally fabricated by screen-printing or paste deposition, followed by a sintering step. Their inherently porous microstructure promotes gas diffusion throughout the film and maximizes the available active surface area. However, this advantage is often accompanied by relatively slower response and recovery times.
- **Thin films** ($\approx 10 - 500$ nm) are typically produced using techniques such as sputtering, evaporation, chemical vapor deposition (CVD), or solution-based methods including sol–gel and spin coating. Owing to their reduced thickness, thin films allow precise control over microstructure and composition, often resulting in higher sensitivity and faster sensor dynamics with respect to thick films of the same material and microstructure. These benefits may, however, come at the expense of increased fabrication complexity, higher costs, and, in some cases, reduced long-term stability.

Since gas sensing is intrinsically a surface-driven phenomenon, the engineering of nanostructures is of paramount importance. Depending on the synthesis route and processing conditions, a wide variety of morphologies can be achieved, including

Table 8.1 Features of thick and thin film sensors

Feature	Thickness	Fabrication	Morphology	Advantages	Limitations
Thick film	1–50 μm	Screen-printing Paste deposition Sintering	Porous Less uniform	High surface area Robust Cost-effective	Slower response Slower recovery Less control over structure
Thin film	10–500 nm	Sputtering Evaporation CVD Sol–gel Spin coating	Compact, highly controlled, nanostructured	High sensitivity Fast response Tunable properties	Higher fabrication cost Possible stability issues

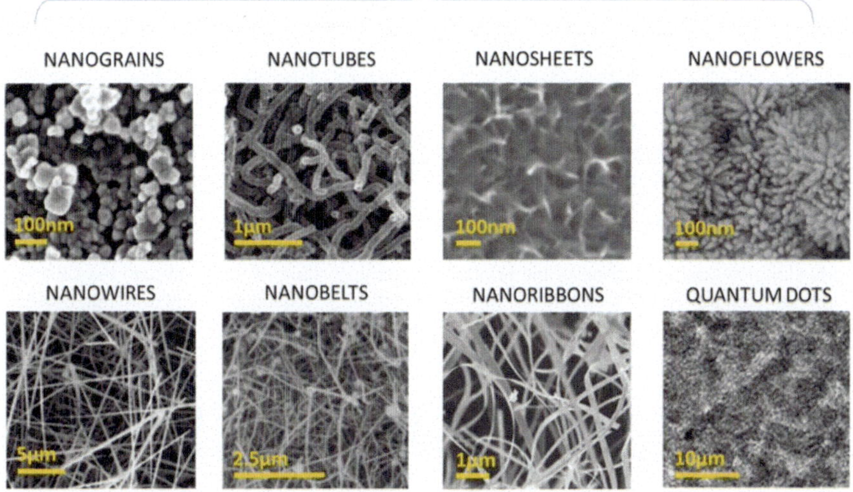

Fig. 8.1 Main chemoresistive sensor nanostructures. All the SEM images are taken from SnO_2 sensor films reproduced from Zonta et al., 2023 under the CC BY 4.0 license [1]

nanoparticles, nanotubes, nanosheets, nanowires, nanobelts, nanoflowers, nanoribbons, and even quantum dots, as illustrated in Fig. 8.1. These geometrical configurations strongly influence both the density of active surface sites and the charge-carrier transport pathways, thereby exerting a decisive impact on sensor performance. Moreover, sensing layers can be realized as thick or thin films, or even deposited onto flexible substrates, opening new perspectives for the development of wearable and portable sensing platforms.

The fabrication of a thick-film gas sensor involves a sequence of well-defined steps, schematically illustrated in Fig. 8.2, which are discussed in detail throughout this Chapter. These steps include: synthesis of the semiconducting powder (see Sect. 8.2); preparation of the printable paste; deposition of the paste onto the substrate by screen-printing; thermal treatments (drying and firing); sensor assembly through bonding techniques; and final laboratory testing.

As a representative example, a typical chemoresistive sensor developed at the SL is schematized in Fig. 8.3. The device consists of a square thick-film sensing layer with an area of approximately $1 \times 1\,\text{mm}^2$ and a thickness of about 30 μm, screen-printed onto an alumina substrate measuring $2.5 \times 2.5\,\text{mm}^2$. The substrate usually hosts interdigitated (comb-shaped) gold electrodes on the front side and a resistive heater, commonly a platinum meander, on the backside to control the operating temperature. The completed sensors are subsequently wire-bonded to TO-39 supports using fine gold wires.

The **substrate** used in gas sensor fabrication fulfills a dual role: it provides mechanical support for the sensing layer and simultaneously acts as an electrical

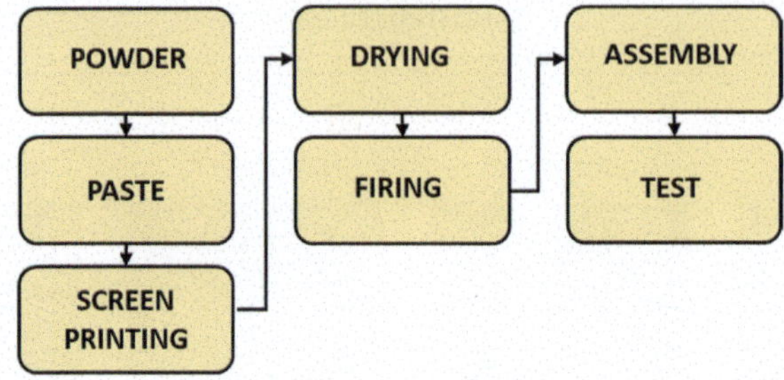

Fig. 8.2 Thick film gas sensors production process scheme that includes: synthesis of the semiconducting powder; preparation of the printable paste; deposition of the paste onto the substrate by screen-printing; thermal treatments (drying and firing); sensor assembly through bonding techniques; and final laboratory testing

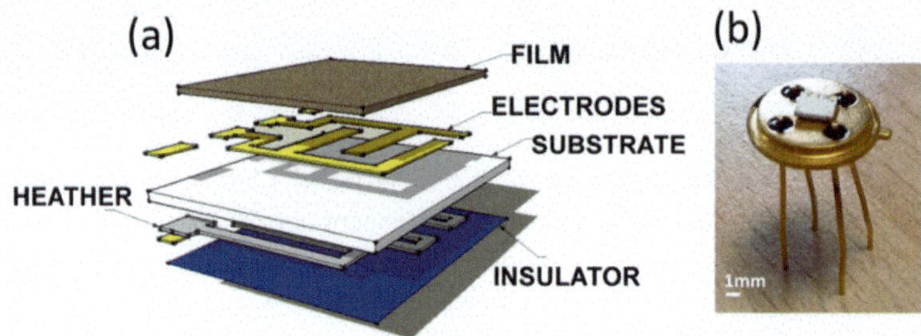

Fig. 8.3 **a** Exploded view of a chemoresistive sensor fabricated at the SL; **b** photograph of an SL sensor. Reproduced from Zonta et al., Chemosensors 2023, 11, 519, under the CC BY 4.0 license

insulator between the functional film and the electronic circuitry. Among the materials commonly employed, alumina (Al_2O_3) is by far the most widely used, owing to its excellent thermal stability, resistance to corrosion and wear, superior dielectric properties, and relatively low cost. Alumina is derived from bauxite, the primary natural source of aluminum. To achieve the desired physicochemical characteristics, alumina powder is finely ground together with other oxides, such as SiO_2, Na_2O, and MgO. This grinding process, typically carried out using a bladed milling system, ensures a homogeneous mixture. The resulting compound is then pressed at different pressures, depending on the target substrate thickness (above or below 1 mm), and subsequently fired for $12 - 24$ h. Ideally, the substrate must remain chemically inert with respect to the sensing reactions and should not actively participate in the detection mechanism. Nevertheless, a certain degree of compatibility between the sensing film and the substrate is essential to avoid mechanical stresses caused by mismatched thermal expansion. For this reason, substrates must possess a smooth and uniform surface with a minimal number of visible defects. Furthermore, the substrate material must exhibit both chemical and physical compatibility with all other materials in contact during sensor fabrication and operation.

To enable deposition onto the substrate and electrical interfacing with the control electronics, the semiconducting active material must be processed into a **paste**. This paste typically consists of three main components:

- the active material;
- the frit;
- organic vehicles.

The active material consists of a nanostructured semiconducting powder with an average particle size typically below 100 nm. Its chemical composition is selected according to the target gas to be detected. The fabrication process begins with the synthesis of this powder in the laboratory, for instance using sol–gel technique, which will be discussed later. In a subsequent step, the powder is dispersed into one or more organic vehicles in carefully controlled proportions to form an ink. A small amount of frit is then added, depending on the nature of the active material. The frit is a glassy oxide mixture based on silica (SiO_2), often modified with alkaline-earth or group-IV oxides to tailor its thermal properties. It serves a dual function: (i) promoting adhesion between particles, thereby enabling the formation of a continuous particle network and the establishment of intergranular potential barriers (often referred to as Schottky-type barriers); and (ii) anchoring the sensing film firmly to the substrate, preventing detachment. The organic vehicles are typically organic solvents containing resins and surfactants, whose role is to ensure proper particle dispersion and to tune the viscoelastic properties of the paste, thus achieving suitable rheological behavior for printing. These organic constituents are completely removed during subsequent thermal treatments. A schematic of the paste preparation process is shown in Fig. 8.4.

Deposition of the sensing paste onto the substrate is most commonly performed using a serigraphic (**screen-printing**) technique, employing a dedicated screen-printing machine (Fig. 8.5). This system includes two key elements:

- the screen, which defines the geometry of the printed pattern;
- the squeegee, a spatula-like tool that applies pressure to force the paste through the screen mesh and deposit it onto the substrate.

A standard screen is composed of a network of finely twisted steel wires, approximately 100 μm in diameter, stretched over an aluminum frame. The spacing between the wires forms apertures, whose density is quantified by the mesh number, defined as the number of wires per linear inch. This parameter is adjusted according to

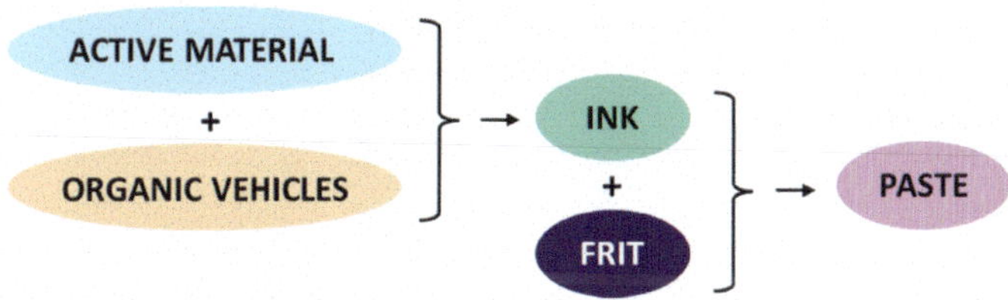

Fig. 8.4 Paste preparation process

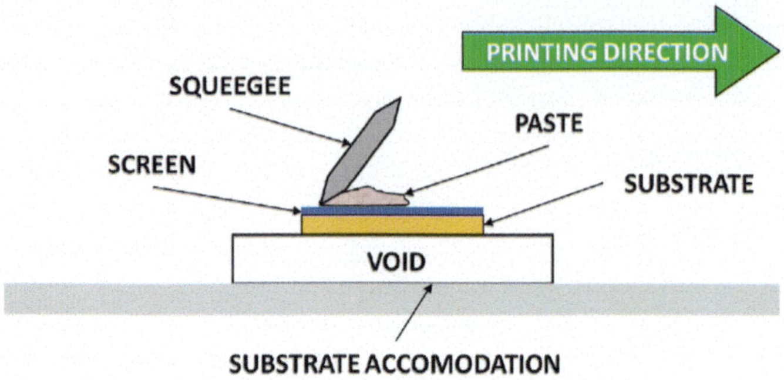

Fig. 8.5 Schematic representation of screen-printing technique

the desired application. The screen surface is coated with a UV-sensitive emulsion onto which the desired pattern is photographically transferred. The regions without emulsion allow the paste to pass through and be printed onto the substrate. The screen therefore has a dual function: it defines the geometry of the printed film and regulates the amount of paste deposited. For optimal performance, the paste must behave as a pseudoplastic fluid, meaning that its viscosity decreases with increasing applied stress, thus facilitating its transfer through the screen. The squeegees must be resistant to the solvents present in the paste. In a typical setup, two squeegees are employed: the first drives the paste through the screen to form the printed pattern (as the one illustrated in Fig. 8.4), while the second redistributes the remaining paste in the opposite direction, resetting it for the next printing cycle.

An example of a sensing paste deposited by screen-printing onto an alumina substrate is shown in Fig. 8.6.

After printing, thermal treatments are essential to ensure the long-term stability and reliability of the sensing film. These treatments are generally divided into two stages, depending on the applied temperature:

Fig. 8.6 Thick-film sensor printed onto an alumina substrate

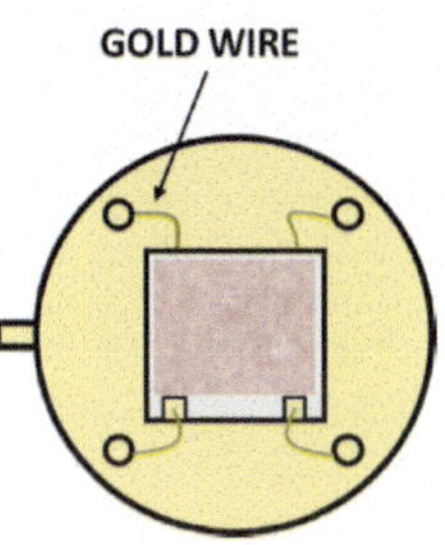

Fig. 8.7 Connection of the sensor to the four-pins on the support using gold wires ($\varnothing \approx 0.06$ mm) bonded by thermo-compression to the heater and the sensing film

- **drying** ($100 - 150\,°C$) aimed at removing volatile solvents;
- **firing** (up to $850\,°C$) intended to eliminate organic additives and stabilize the nanostructure.

The typical operating temperature of chemoresistive gas sensors lies in the range $300 - 500\,°C$. Within this interval, thermally activated charge transport is enhanced while excessive grain coalescence, which would degrade sensor performance, is avoided. The firing step is particularly important for stabilizing the grain size: if the operating temperature is lower than the firing temperature, the nanostructure remains stable during operation. In applications where thermal activation is impractical, photoactivation provides an alternative strategy. By illuminating the sensing material with light of suitable wavelengths, additional active sites can be generated, thereby enhancing the sensor response. As will be discussed later, only certain semiconductors (e.g., WO_3, ZnO, CdS, SnS_2) exhibit both chemoresistive and photoconductive behavior, enabling effective exploitation of this mechanism.

The final assembly of the sensor is carried out through a **bonding process**, which consists of connecting the sensing element to a supporting structure via four pins, as illustrated in Fig. 8.7. These pins are joined to the integrated heater by thermo-compression welding and simultaneously connected to the contact pads on the substrate supporting the sensing film. Electrical connections are realized using high-purity gold wires (99.99%) with a diameter of approximately 0.06 mm.

This delicate operation is carried out with the aid of a dedicated apparatus, referred to as a bonding machine (Fig. 8.8).

8.2 Semiconductor Synthesis

A variety of methods can be employed for the synthesis of semiconductor nanostructures. Among the most widely adopted are the **hydrothermal method**, which involves aqueous-based synthesis in a sealed and pressurized vessel, allowing precise control over nanostructure growth, and the **sol–gel method**, here described and schematically summarized in Fig. 8.9. This approach (routinely employed by the SL at the University of Ferrara) has demonstrated particular effectiveness in the controlled fabrication of sensing materials. The sol–gel method can be employed

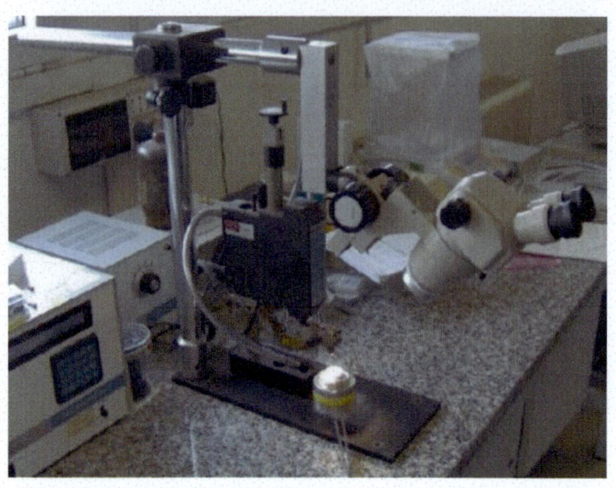

Fig. 8.8 Bonding machine at the SL

for the preparation of both pure or multicomponent metal oxides by hydrolysis of inorganic salts in water or of metal–organic compounds in non-aqueous solutions. The process begins with the preparation of a solution containing metal-based starting reagents (typically metal alkoxides or metal salts) dissolved in either water or an alcoholic solvent. In the next step, the system undergoes *hydrolysis* and *condensation* reactions. During hydrolysis, the metal precursors react with water, leading to the formation of hydroxyl groups (–OH). These groups then undergo condensation reactions, forming metal–oxygen-metal (M–O–M) bonds and releasing small molecules like water or alcohol. This step leads to the formation of a colloidal suspension, known as "**sol**", where small oxide nanoparticles are dispersed in the liquid medium. As the sol ages, it gradually transforms into a "**gel**", a three-dimensional porous network that traps the solvent within its structure. This occurs through further condensation and crosslinking of the nanoparticles. The next stage involves the removal of the solvent, typically through *evaporation*, which results in the drying of the gel and the formation of a solid material. To stabilize the structure and improve crystallinity, the dried gel undergoes a process called *calcination*, which consists of irreversible thermal oxidation at high temperatures (usually between 400 and 900 °C). This step eliminates residual organic components and fully converts the material into the desired metal oxide phase with improved structural and chemical stability.

The final product is a nanostructured metal oxide, often characterized by high surface area, porosity, and sensitivity, properties that are crucial for efficient gas sensing in chemoresistive devices.

On the other hand, the hydrothermal method is a widely adopted technique for the synthesis of metal oxide nanostructures tailored for chemoresistive sensor applications. It involves carrying out heterogeneous chemical reactions in the presence of a solvent (either aqueous or non-aqueous) at temperatures above room temperature and under pressures greater than 1 atm, all within a closed system such as an autoclave [2]. This environment enables the controlled nucleation and growth of

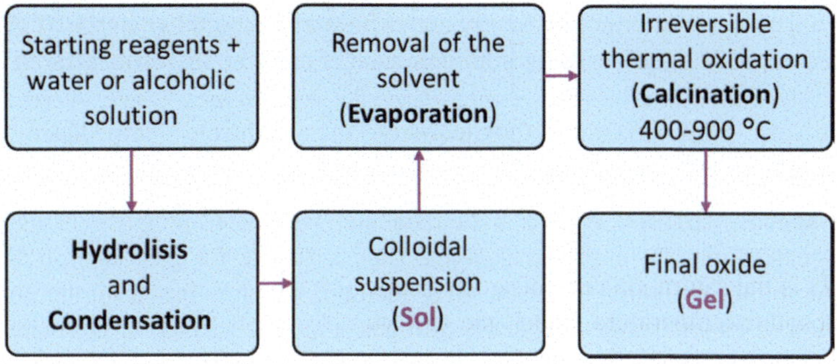

Fig. 8.9 Sol–gel process schematized in its main steps

crystalline oxide phases at relatively low temperatures compared to traditional solid-state synthesis. One of the key advantages of the hydrothermal process is its ability to produce materials with well-defined morphologies, such as nanorods, nanowires, or hierarchical structures, which offer high surface-to-volume ratios and improved gas diffusion pathways: both critical factors for enhancing the sensitivity and selectivity of chemoresistive gas sensors. Additionally, the hydrothermal method often leads to high-purity products with fewer defects, which contributes to better electronic properties and stability in sensing performance.

8.3 Characterization Techniques for Chemoresistive Materials

The development of efficient chemoresistive gas sensors critically depends on a deep understanding of the structural, morphological, thermal, optical, and electrical properties of the sensing materials, which are typically synthesized in the form of nanostructured powders or thin/thick films. An appropriate and systematic characterization is therefore essential to establish clear correlations between material properties and sensor performance. The characterization techniques commonly employed can be broadly divided into two main categories: (i) structural and morphological characterization of powders and films, and (ii) functional tests aimed at evaluating material behavior under specific external stimuli.

The first category includes the following characterization methods:

- specific surface area (Brunauer–Emmett–Teller method, BET);
- grain size (transmission electron microscopy, TEM);
- crystal structure (X-ray diffraction, XRD);
- morphology (scanning electron microscopy, SEM; TEM; atomic force microscopy, AFM).

The second category comprises functional tests, namely:

- thermal analysis (thermogravimetry and differential thermal analysis, TG/DTA)
- optical characterization (photoluminescence; Fourier transform infrared spectroscopy, FTIR)
- electrical characterization (conductivity measurements, impedance spectroscopy).

The **specific surface area** is a fundamental parameter in gas sensing, as it directly determines the number of active sites available for gas adsorption. It is most commonly evaluated using the BET method, which is based on the physical adsorption of gas molecules, typically nitrogen, onto the surface of the material at cryogenic temperature ($-196\,°C$,, corresponding to liquid nitrogen). A gas sorption analyzer measures the quantity of adsorbed gas, and the BET equation is then applied to calculate the specific surface area, expressed in m^2/g. A high specific surface area is particularly advantageous for chemoresistive sensors, as it enhances sensitivity by promoting more efficient interaction between the sensing material and the gaseous species.

Grain size and crystallite dimensions at the nanoscale are investigated by TEM. In this technique, a high-energy electron beam is transmitted through an ultra-thin specimen, and the resulting contrast and diffraction patterns provide detailed information on particle size, shape, crystallinity, and internal structure. TEM enables direct observation of the particle size distribution and the degree of agglomeration, both of which strongly influence the electrical transport properties and the sensing performance of the material.

XRD represents the primary technique for determining the **crystal structure** and phase composition of a material. When X-rays interact with a crystalline sample, they are diffracted according to Bragg's law, generating characteristic diffraction patterns that depend on the atomic arrangement within the lattice. These patterns are recorded using an X-ray diffractometer and allow identification of the crystalline phases present. Moreover, the average crystallite size can be estimated from peak broadening using the Scherrer equation, making XRD an indispensable tool for verifying the formation of the desired metal oxide phase following synthesis or thermal treatment.

Surface **morphology** plays a crucial role in gas sensing and is examined using several complementary techniques. SEM provides high-resolution images of surface topography by scanning the sample with a focused electron beam and detecting secondary or backscattered electrons. SEM images reveal features such as grain size, porosity, film continuity, and uniformity. TEM, in addition to its role in grain size analysis, also contributes to morphological characterization by offering insight into internal structures at nanometer or sub-nanometer resolution. AFM is a non-destructive technique that employs a nanometric tip mounted on a cantilever to scan the surface. The interaction forces between the tip and the sample are recorded to generate three-dimensional surface maps with atomic-scale resolution in ideal conditions. AFM is particularly useful for assessing surface roughness and texture, especially in thin films deposited on substrates.

Thermal analysis techniques, such as thermogravimetric analysis (TGA) and differential thermal analysis (DTA), are used to investigate the thermal stability, mass loss processes, and phase transformations of sensing materials. TGA measures changes in sample mass as a function of temperature under a controlled atmosphere (e.g., air or nitrogen) and is valuable for monitoring solvent evaporation, decomposition of precursors, or the removal of residual organic components after synthesis. DTA records the temperature difference between the sample and an inert reference during heating, allowing detection of endothermic and exothermic events such as crystallization, oxidation, or phase transitions. These analyses are performed using thermal analyzers equipped with precision balances and programmable furnaces and are essential for optimizing calcination and firing conditions.

Optical characterization employs techniques like photoluminescence (PL) spectroscopy, that involves excitation of the material with ultraviolet or visible radiation, followed by analysis of the emitted light. The resulting emission spectra provides information on electronic defects, surface states, and vacancies (particularly oxygen vacancies) which play a central rôle in the sensing mechanisms of metal oxide semiconductors. On the other hand, FTIR is employed to identify chemical bonds and functional groups by analyzing the absorption of infrared radiation at characteristic wavelengths. FTIR is especially useful for confirming the removal of organic precursors, as well as detecting hydroxyl groups or other surface species that may influence gas adsorption and reaction kinetics. Both PL and FTIR measurements are typically carried out using dedicated spectrophotometers and can be applied to both powder and films.

Electrical characterization is fundamental for evaluating the chemoresistive response of sensing materials. The sensing mechanism is based on changes in electrical resistance induced by surface reactions between adsorbed oxygen species and target gas molecules. Electrical conductivity is commonly measured using two-probe or four-probe configurations, or through interdigitated electrodes patterned on substrates. Impedance spectroscopy provides more detailed insight into the electrical behavior by probing the response of the material to an alternating current over a broad frequency range. This technique allows separation of contributions from grains, grain boundaries, and electrode interfaces, which are typically modeled using equivalent electrical circuits. The resulting Nyquist or Bode plots offer valuable information on charge transport mechanisms and the effects of gas adsorption. These measurements are performed using precision LCR meters or impedance analyzers, often under controlled temperature and gas atmospheres.

8.4 Main Sensing Materials

Tin dioxide (SnO_2) is one of the most investigated and technologically mature semiconductor oxides for chemoresistive gas sensing applications [3–21]. As an n-type wide-bandgap semiconductor ($E_g \approx 3.6\,\text{eV}$), SnO_2 exhibits a strong dependence of its electrical conductivity on the surrounding gaseous atmosphere, making it particularly effective for the detection of both oxidizing and reducing gases. Its remarkable gas-sensing capability originates from surface-controlled charge transfer processes, where adsorbed oxygen species interact with target gases, modulating the width of the electron depletion layer. SnO_2 has been extensively employed since the development of the first commercial Taguchi-type gas sensor in 1970 [3]. Despite its robustness, chemical stability, and high sensitivity, SnO_2 suffers from limited selectivity and humidity interference, which remain critical challenges for environmental and industrial monitoring applications. From an electronic standpoint, SnO_2 exhibits two donor levels located at $E_{d1} = 0.034$ eV and $E_{d2} = 0.145\,\text{eV}$ below the conduction band. These shallow donors, typically associated with oxygen vacancies and interstitial tin, are responsible for SnO_2 intrinsic n-type conductivity. The material displays different responses depending on the type of target gas: oxidizing gases (e.g., NO_2) enhance the electron depletion layer and increase resistance, while reducing gases (e.g., CO, CH_4, H_2) decrease resistance through electron donation. The sensing mechanism is also temperature-dependent and oxidizing gases are preferentially detected at lower operating temperatures, whereas reducing gases require higher temperatures for efficient surface reactions. To enhance selectivity and sensitivity, numerous strategies have been explored, primarily involving doping and surface modification. Noble metal additions, such as gold (Au) and palladium (Pd), have been shown to improve response and lower the optimal operating temperature. This enhancement is attributed to spillover effects, catalytic promotion of gas dissociation, and the formation of Schottky barriers that modulate charge carrier dynamics. In particular, Chiorino et al. (1997) demonstrated that Pd-loaded SnO_2 films exhibit significantly higher responses to oxidizing gases and reduced working temperatures compared to undoped SnO_2 [4]. Beyond noble metals, ionic dopants introduced into the crystal lattice also play a major role in tuning gas sensitivity. For instance, Mo and W dopants suppress responses to reducing gases such as CO, while enhancing those toward oxidizing species like NO_2. Conversely, V and La dopants confer higher sensitivity to gases such as SO_2 and CO_2. These modifications not only affect the surface reactivity but also alter the electronic structure by introducing donor or acceptor levels that influence carrier concentration and oxygen vacancy distribution. The synthesis route strongly determines the morphological and electronic properties of SnO_2. Nanostructured powders prepared by sol–gel, hydrothermal, or precipitation methods exhibit tunable grain sizes and surface area, both of which critically influence gas-sensing performance. When the grain size approaches the Debye length (typically below 10 nm), the entire particle becomes depleted under ambient air, resulting in a maximized modulation of resistance upon gas exposure. Experimental evidence confirms that reducing grain dimensions to the order of depletion width enhances gas response, though excessively small grains can induce high baseline resistance and slow recovery. Conversely (as the sample in Fig. 8.10), firing at high temperatures ($\approx 850\,°\text{C}$ for 1h) promotes grain coalescence (up to $20 - 30$ nm), improving crystallinity and mechanical stability but reducing surface area and, consequently, sensitivity. A major limitation of SnO_2-based sensors remains their cross-sensitivity to humidity. Even small amounts of water vapor can significantly influence electrical resistance due to competitive adsorption between

Fig. 8.10 SnO_2 film, fired for 1 h at 850 °C with average grain size > 20 nm

oxygen species and hydroxyl groups. Malagù et al. (2004) demonstrated that this phenomenon becomes particularly relevant when the depletion layer is thinner than 20 nm, suggesting that sensing films should be composed of nanoparticles smaller than 10 nm and operated at moderate temperatures to achieve optimal stability [5]. Despite these challenges, SnO_2 remains the most widely used material for gas sensors due to its versatility, affordability, and well-established processing routes. Through controlled morphology, narrow grain size distribution, and targeted doping with noble metals or transition ions, it is possible to achieve improved selectivity and stability toward specific gases such as NO_2. The combination of these factors makes SnO_2 a reference material and a benchmark for the design of advanced nanostructured chemoresistive sensors.

Titanium Dioxide (TiO_2) is another important material for gas sensing; it is a widely employed n-type semiconductor for chemoresistive gas sensors fabrication, owing to its favorable physical and chemical properties: it is low cost, non-toxic, chemically stable under harsh atmospheres, and can be synthesized by many wet-chemical and vapor-phase routes [22–31]. TiO_2 exists in several polymorphs (most notably anatase, rutile, and brookite) of which anatase and rutile are the most relevant for gas sensing. The band gap is approximately 3.2 eV for anatase and $\sim 3.0-3.05$ eV for rutile, with anatase showing higher photoactivity under UV excitation and rutile being the more stable phase at elevated temperatures. The relative proportions of anatase vs rutile strongly affect surface adsorption, electron affinity, work function, and thereby the gas sensing behavior. In TiO_2-based chemoresistive gas sensors, morphology and grain size critically affect sensitivity, selectivity, and stability: well-defined and uniform morphologies (e.g., equiaxed nanoparticles, nanorods, nanosheets) afford reproducible surface characteristics and reduce inactive or "dead" facets, though careful synthetic control is required to avoid irregular shapes. A narrow grain-size distribution optimizes the sensor behavior: when grains approach sizes comparable to twice the Debye length, full depletion in air leads to maximized resistance modulation under gas exposure. Reducing the crystallite size toward the few-nanometer range often enhances sensitivity, though overly small grains increase baseline resistance, noise, and slow recovery. Thermal annealing (e.g., firing or annealing at high temperatures) improves crystallinity and stability but also tends to increase grain size and reduce surface area, potentially degrading sensitivity; the annealing atmosphere further modulates oxygen vacancy concentration and defect states, influencing baseline conductivity and response to oxidizing vs reducing gases. Doping and surface modification (e.g., Al doping, Cr doping, heterojunctions, noble metal decoration) serve to tailor electronic structure, control defects, provide catalytic activation, and thereby improve selectivity (for instance towards NO_2), lower operating temperature, and stabilize response. Thus, for TiO_2 sensors one must balance morphology control, optimal grain size, thermal treatment, and dopant engineering to reach high sensitivity, reproducibility, and selective detection in realistic conditions. TiO_2-based materials find applications in several technological sectors. In particular, they are used in automotive exhaust sensors for monitoring oxygen partial pressure and controlling the air/fuel ratio, and in the detection of atmospheric pollutants, where TiO_2 offers chemical stability and reproducible operation under varying environmental conditions. These sensors exhibit stable responses only when grain dimensions are in the nanometric range, since large grains suppress surface-controlled processes essential for chemoresistive behavior. When heated, TiO_2 undergoes an anatase-to-rutile phase transition, a process that distinguishes it from SnO_2. In Fig. 8.11, an example of pure TiO_2 in the anatase phase and, in Fig. 8.12, TiO_2 in the rutile phase: it is strongly evident the grain size difference between the two samples. The stabilization of the anatase phase is of significant practical interest, as anatase exhibits superior gas-sensing performance at moderate temperatures. This stabilization can be achieved through the addition of pentavalent ions (e.g., Ta^{5+}, Nb^{5+}), which hinder the transformation to rutile. For example, the incorporation of 10% Ta delays the anatase-to-rutile transition up to 950 °C. Alternatively, the use of the sol–gel method followed by hydrothermal treatment allows maintenance of the anatase phase up to approximately 850 °C without introducing foreign elements. The addition of vanadium ions, instead, facilitates the anatase-to-rutile transformation, whereas tantalum (or niobium) retards it. Interestingly, the simultaneous addition of vanadium and tantalum partially compensates these two opposite effects: while vanadium promotes catalytic activity and enhances the response to oxidizing gases, tantalum reduces the grain coalescence tendency induced by vanadium. The synergy between these dopants yields materials that combine high sensitivity, catalytic efficiency, and morphological stability. A representative study by Carotta et al. (2005) demonstrated the preparation of TiO_2 doped with both Ta and V (a sensor named **TiTaV**), showing excellent performance and long-term stability for gas sensing [32–34]. From a theoretical perspective, TiO_2 gas-sensing behavior can be described by applying the Schottky relation to a spherical geometry, in cases where the depletion approximation fails. This approach highlights the strong grain-size dependence of TiO_2 sensing properties, governed by the variation of the surface state density with the grain radius. As a result, smaller TiO_2 grains exhibit a more pronounced modulation of conductivity in response to gas adsorption. In contrast, SnO_2 often shows a weaker dependence on particle size due to the near constancy of N_S, a phenomenon confirmed experimentally. Moreover, TiO_2 displays much smaller electrical conductivity than SnO_2, reflecting the deeper position of its donor-like defects within the bandgap. By comparing TiO_2 and SnO_2, it emerges that TiO_2 is less reactive toward reducing gases. However, both materials share the rutile structure, enabling the formation of SnO_2-TiO_2 solid solutions that exploit the complementary electronic and catalytic properties of each oxide for enhanced gas-sensing performance, particularly in thick-film devices [35–40].

(Ti,Sn)O_2 solid solutions form readily because Ti^{4+} and Sn^{4+} can occupy the same octahedral sites of the rutile-type lattice, producing a continuous series of compositions $Ti_xSn_{1-x}O_2$ whose X-ray diffraction peaks (for example the (211) reflection) shift quasi-linearly with Ti content. Carotta et al. (2009) identified $x \approx 0.2$ as a compositional boundary where the material's structural and transport behaviour changes from SnO_2-like to TiO_2-like [35]. Electronically, these mixed oxides retain n-type conduction but display transport and sensing signatures that are not a simple interpolation of the end-members: Arrhenius plots of conductivity for Ti-rich compositions lack the characteristic S-shaped feature often observed in pure SnO_2 (a manifestation of complex surface reaction kinetics and thermally activated

Fig. 8.11 Pure TiO_2 film, fired for 1 h at 650 °C in the anatase phase

Fig. 8.12 Pure TiO_2 film, fired for 1 h at 850 °C in the rutile phase

charge-trapping phenomena), indicating altered surface reaction pathways and activation energies in the solid solution as shown by Schipani et al. (2012) [36]. The chemisorbed oxygen chemistry also changes across the series: the $O_2 \rightarrow O^-$ (and related $O^- \rightarrow O^{2-}$) surface charge-transfer transitions that are prominent in SnO_2 become apparent in the $\sim 200 - 400\,°C$ window, whereas in TiO_2 these transformations are less visible at the same temperatures because the relevant bulk donor states in TiO_2 lie deeper in the band gap and are not thermally ionized under the same conditions; consequently, Ti–rich compositions tend to show weaker temperature-activated surface donor ionization and different sensitivity/humidity cross-sensitivity profiles. From a practical point of view for thick-film gas sensors, $(Ti,Sn)O_2$ solid solutions combine the high chemical reactivity of SnO_2 with the lower humidity cross-sensitivity and improved thermal stability of TiO_2, and, through compositional tuning (notably near the $x \approx 0.2 - 0.3$ range), controlled defect engineering and appropriate annealing, one can tailor oxygen vacancy concentrations, activation energies and surface adsorption sites to optimize selectivity and operating temperature for target analytes.

Tungsten trioxide (WO_3) is a versatile n-type semiconductor oxide extensively studied for its multifunctional applications in electrochemical devices, photocatalysis under both UV and visible light, and particularly in chemoresistive gas sensing [41–51]. Its high chemical stability, tunable stoichiometry, and rich polymorphism make it one of the most adaptable materials among transition metal oxides. WO_3 exhibits a band gap ranging between 2.5 and 2.8 eV, depending on its crystallographic phase and synthesis route, allowing partial photoactivation under visible light, a feature that extends its use to photo-assisted sensing and catalytic devices. Structurally, WO_3 possesses a wide variety of crystalline forms (monoclinic, triclinic, orthorhombic, and hexagonal) which can interconvert depending on temperature and oxygen stoichiometry. This structural flexibility leads to significant variations in electronic and surface properties, especially in the concentration of oxygen vacancies that act as donor states, providing n-type conductivity. The presence of such vacancies also facilitates adsorption–desorption dynamics of oxygen species, a key process for gas sensing. In fact, despite differences in morphology and synthesis route, the sensing response of WO_3 is governed primarily by its surface oxygen vacancies, which act as the main reactive sites toward analyte gases, thus explaining the remarkable uniformity observed among the various WO_3-based sensor. Notably, WO_3 can exhibit conductance up to three orders of magnitude higher than that of SnO_2, due to its narrower band gap and higher intrinsic carrier concentration. From an electronic standpoint, WO_3 displays three distinct conduction regimes as a function of temperature, analogous to those observed in SnO_2, which reflect the progressive activation of surface states, bulk conduction, and intrinsic semiconductor behavior. The energy barrier height exhibits a similar temperature trend to that of SnO_2, though with a markedly lower absolute value for WO_3, implying faster surface reactions and lower activation energy for charge transfer. These properties make WO_3 particularly sensitive to oxidizing gases such as NO, NO_2, H_2S, and NH_3, for which it exhibits strong and stable responses at moderate to high operating temperatures. Despite its excellent response to oxidizing gases, relatively few studies have explored the detection of reducing gases (e.g., volatile organic compounds-VOCs and hydrocarbons) using WO_3. This asymmetry arises from the inherent electronic configuration of the material: its surface tends to favor the adsorption of electron-withdrawing species, while the catalytic dissociation of reducing molecules is less efficient without suitable dopants or heterostructural modifications. The introduction of noble metals (e.g., Pt, Pd, Au) or other transition metal oxides can improve selectivity toward reducing gases by creating additional reaction sites and modulating the local Fermi level. One of the major challenges associated with WO_3 lies in its pronounced grain coalescence during thermal treatments, particularly when assisted by phase transitions at elevated temperatures, as shown in Fig. 8.13. WO_3 undergoes exaggerated grain growth beyond 30 nm when fired above 700 °C, leading to a substantial reduction in surface area and, consequently, in sensitivity. Unfortunately, high-temperature processing is necessary to achieve the desired crystalline stability and phase purity. This trade-off between structural stability and specific surface area is a limiting factor for practical sensor fabrication, as nanoparticles smaller than $\sim$ 30 nm are not thermally stable under the conditions required for device consolidation.

$(Sn,W)O_3$ solid solutions have been investigated to mitigate grain growth and enhance morphological control. The addition of $1-5\%$ Sn into the WO_3 lattice progressively modifies its electronic properties, shifting sensing behavior from reducing-gas detection (typical of SnO_2) towards oxidizing-gas detection (characteristic of WO_3) [52, 53]. However, a solid solution containing 10% Sn does not significantly improve the response of WO_3 to oxidizing gases, indicating that excessive Sn substitution may disturb the WO_3 lattice without providing additional catalytic benefit. Morphological studies by SEM revealed that the addition of Sn markedly suppresses grain coalescence, allowing the preservation of smaller, more uniform grains even after firing at 750 °C. This structural stabilization explains the improved reproducibility and slower aging of $(Sn,W)O_3$-based films. Comparative studies on the sensing performance of pure and Sn-doped WO_3 films further illustrate the influence of dopant concentration and thermal treatment. For instance, the response magnitude to 10 ppb of NO_2 was found to be highest for pure WO_3 fired at 750 °C, while $(Sn,W)O_3$ (Sn: 10%) samples exhibited lower responses at the same temperature but improved morphological control and stability at higher firing temperatures ($650-850$ °C). This behavior highlights the delicate balance between grain size, crystallinity, and surface chemistry in determining sensor performance. In summary, WO_3-based gas sensors exhibit exceptional sensitivity and selectivity toward oxidizing gases, owing to their high oxygen vacancy concentration, favorable electronic structure, and tunable morphology. However, the material's tendency toward excessive grain growth at elevated temperatures poses challenges for achieving nanoscale control and long-term stability. Doping strategies, particularly through the formation of $(Sn,W)O_3$ solid solutions, provide a promising route to tailor both morphology and selectivity, ensuring reliable detection of key oxidizing pollutants such as NO_2 at trace levels.

Zinc oxide (ZnO) is a direct wide band gap ($E_g = 3.37$ eV) n-type semiconductor with a high exciton binding energy of approximately 60 meV and strong near-band-edge ultraviolet (UV) emission around 380 nm [54–68]. These intrinsic properties,

Fig. 8.13 Grain coalescence of WO_3 when fired at 650 °C (top panel) and at 850 °C (bottom panel), respectively

combined with its chemical stability, abundance, and ability to form various nanostructures, make ZnO a highly versatile material for optoelectronic, photocatalytic, and sensing applications. ZnO can be easily synthesized in different nanoforms (including nanoparticles, nanowires, nanobelts, nanorods, and nanotetrapods) through a range of solution-based and vapor-phase techniques, which enables precise tailoring of its morphology and surface area for specific applications. In Fig. 8.14 a ZnO sample synthesized in the shape of nanoparticles. Historically, ZnO has been one of the earliest materials investigated for chemoresistive gas sensors, showing good response toward oxidizing gases such as O_3, NO_2, and O_2, as well as reducing gases including H_2, H_2S, LPG, formaldehyde (HCHO), and NH_3. However, traditional ZnO-based sensors typically required high working temperatures ($400 - 500$ °C) to achieve sufficient surface reactivity, leading to high power consumption and thermal instability over time. Furthermore, the material has been characterized by poor selectivity, responding simultaneously to multiple gas species, a feature that has limited its widespread adoption compared to SnO_2 and WO_3. However, in recent years, with the advent of advanced powder processing techniques, particularly hydrothermal synthesis and nanoceramic processing, ZnO has experienced renewed interest as a sensing material. These methods allow for improved control over particle size, morphology, and defect distribution, thereby enhancing sensor performance. For instance, Baruwati et al. (2006) reported the hydrothermal synthesis of ZnO nanoparticles in double-distilled water at 120 °C under autoclave conditions, producing materials with a unimodal particle-size distribution (average diameter $\approx$ 30 nm) [59]. The hydrothermal treatment was found to hinder grain growth compared to sol–gel routes prepared under similar precursor conditions, resulting in higher surface area and improved gas-sensing properties. Such morphological control is crucial, since the sensor's electrical response depends strongly on the surface-to-volume ratio and the accessibility of adsorption sites. From a structural and electronic standpoint, the wurtzite crystal structure of ZnO accommodates oxygen vacancies and zinc interstitials that act as donor states, conferring its n-type conductivity. These intrinsic defects play a pivotal role in gas sensing, serving as active sites for adsorption and redox reactions with target gases. In oxidizing environments, adsorbed oxygen species extract electrons from the conduction band, increasing the material's resistance, whereas exposure to reducing gases re-injects electrons, decreasing resistance. The magnitude and kinetics of this modulation depend on grain size, defect density, and surface chemistry. As with other semiconducting oxides, when the grain size becomes comparable to twice the Debye length, the entire grain may be depleted, maximizing sensitivity. ZnO-based sensors show strong affinity for oxidizing gases, particularly ozone (O_3), toward which they exhibit rapid and reproducible responses. However, selectivity remains a key challenge, as cross-sensitivity to humidity and other oxidizing species is frequently observed. Despite this, ZnO remains attractive for ozone monitoring applications due to its fast response and recovery characteristics, especially when the material is synthesized with uniform nanoscale morphology. One of the most promising developments in ZnO sensing technology is the use of photoactivation to overcome the limitations imposed by high operating temperatures. The integration of UV light-emitting diodes (LEDs) as excitation sources allows ZnO-based sensors to operate effectively at or near room temperature. Under UV irradiation, photogenerated electron–hole pairs enhance surface reactions with adsorbed gases, thereby enabling the detection of oxidizing and reducing species without the need for thermal activation. This photo-assisted gas sensing mechanism significantly reduces power consumption and improves sensor stability. For instance, UV-assisted ZnO sensors have demonstrated excellent ozone detection capabilities at ambient temperature, combining high sensitivity, short response times, and low noise levels. Beyond gas sensing, the multifunctional nature of ZnO has positioned it at the intersection of several technological domains. Its wide range of applications include photocatalysis, piezoelectric energy harvesting, optoelectronics, and photovoltaic conversion. The ability to integrate ZnO nanostructures into hybrid systems

Fig. 8.14 Sample of ZnO nanoparticles

and flexible substrates makes it a candidate material for next-generation smart sensors capable of operating under low-power or self-powered conditions. In summary, ZnO remains a material of great scientific and technological interest for chemoresistive gas sensors. Although its popularity declined due to issues of selectivity and high temperature operation, advances in hydrothermal synthesis, nanostructuring, and photoactivation have revitalized research in this field. By leveraging controlled morphology, defect engineering, and UV-assisted operation, ZnO-based sensors offer a promising route toward efficient, stable, and low-temperature detection of key atmospheric and industrial gases, particularly ozone.

Metal sulfides represent a promising alternative to traditional metal oxides in chemoresistive sensing, offering distinct surface chemistry and narrower bandgaps that can enable lower operating temperatures and enhanced selectivity. Among these, CdS and SnS_2 have been studied for their gas sensing behavior [69–73]. CdS is a direct bandgap semiconductor ($E_g \approx 2.40\,\text{eV}$) known for its piezoelectric, photoconductive, electroluminescent, and chemoresistive functionalities. As a gas sensor, CdS exhibits high selectivity to alcohol vapors, and the amplitude of its response typically peaks under green-light excitation, suggesting a strong coupling between photoexcitation and gas-adsorbate interaction. However, pristine CdS-based sensors are still rare, partly owing to issues of baseline conductivity, stability, and selectivity in complex atmospheres. In Fig. 8.15 a SEM image of CdS. SnS_2, a layered n-type sulfide, has attracted attention for its high selectivity and sensitivities comparable to oxide films used in gas sensing. In standard chemoresistive operation (without external illumination), SnS_2 often show reproducible and stable responses to gases such as aldehydes, ketones, and alcohols (i.e., carbonyl and hydroxyl functional groups) highlighting its potential as a selective sensor material. However, SnS_2 is not particularly suitable for photo-activated chemoresistive sensing using light sources in the 400 – 645 nm region, likely due to insufficient absorption or inefficient photo-carrier generation in that band (Fig. 8.16).

Another relevant reason for which metal sulfides are studied and employed as chemoresistive sensors is because they are more stable in their baseline during

Fig. 8.15 Sample of CdS nanoparticles

Fig. 8.16 Sample of SnS_2 nanoparticles

time than their metal oxide counterparts. The remarkable baseline stability typically observed in metal sulfide-based chemoresistive sensors, such as CdS and SnS_2, originates from their intrinsically limited interaction with atmospheric oxygen and the consequent suppression of oxygen in-diffusion into the lattice. Unlike metal oxides, where oxygen exchange processes (adsorption, desorption, and bulk diffusion) continuously alter the surface stoichiometry and donor density (leading to slow drifts in baseline resistance), metal sulfides possess a denser anionic framework built from larger, less electronegative sulfur atoms that form strong covalent-ionic bonds with the metal cations. This compact structure hinders oxygen incorporation and diffusion significantly lower compared to oxides, thereby maintaining a stable Fermi level and a well-defined depletion layer at the surface. Moreover, in sulfides the adsorption of oxygen or target analytes is largely confined to the outermost surface and does not propagate into the bulk, avoiding long-term modifications of charge-carrier concentration. Since there is no intrinsic lattice oxygen to participate in redox exchange, surface reactions in metal sulfides are rapidly reversible and do not induce slow stoichiometric changes. Consequently, the low oxygen diffusivity and absence of self-exchange dynamics in these materials fundamentally account for their minimal baseline drift and superior long-term electrical stability compared to conventional metal oxide gas sensors.

References

1. Zonta G, Rispoli G, Malagù C, Astolfi M (2023) Overview of gas sensors focusing on chemoresistive ones for cancer detection. Chemosensors 11(10):519
2. Byrappa K, Yoshimura M (2013) Handbook of hydrothermal technology. Elsevier
3. Taguchi N (1970) Gas-detecting element and method of making same. Japanese Patent 45-38200
4. Chiorino A, Ghiotti G, Prinetto F, Carotta MC, Martinelli G, Merli M (1997) Characterization of SnO_2-based gas sensors: a spectroscopic and electrical study of thick films. Sens Actuators B Chem 44:474–482
5. Malagù C, Carotta MC, Fissan H, Guidi V, Kennedy MK, Kruis FE, Martinelli G, Maffeis TGG, Owen GT, Wilks SP (2004) Surface state density decrease in nanostructured polycrystalline SnO_2: modelling and experimental evidence. Sens Actuators B 100:283–286
6. Samson S, Fonstad CG (1973) Defect structure and electronic donor levels in stannic oxide crystals. J Appl Phys 44:4618–4621
7. Korotcenkov G (2007) Metal oxides for solid-state gas sensors: what determines our choice? Mater Sci Eng B 139(1):1–23
8. Comini E, Baratto C, Concina I, Faglia G, Falasconi M, Ferroni M, Galstyan V, Gobbi E, Ponzoni A, Vomiero A, Zappa D, Sberveglieri V, Sberveglieri G (2013) Metal oxide nanoscience and nanotechnology for chemical sensors. Sens Actuators B Chem 179:3–20
9. Nahirniak S, Dontsova TA (2017) Effect of modification/doping on gas sensing properties of SnO_2. Nano Res Appl 3(2)
10. Traversa E (1995) Ceramic sensors for humidity detection: state-of-the-art and future developments. Sens Actuators B Chem 23(2–3):135–156
11. Kohl D (2001) Surface processes in the detection of reducing gases with SnO_2-based devices. J Phys D Appl Phys 34:R125–R149
12. Sberveglieri G (1992) Gas sensors: principles, operation and developments. Springer, Heidelberg
13. Zonta G, Astolfi M, Cerboni N, Gherardi S, Kasprzak M, Malagù C, Steinegger P, Vincenzi D, Chiera NM (2024) Gas-sensing performance of SnO_2-based chemoresistive sensors after irradiation with alpha particles and gamma-rays. J Radioanal Nucl Chem
14. Carotta MC, Dallara C, Martinelli G, Passari L, Camanzi A (1991) CH_4 thick-film gas sensors: characterization method and theoretical explanation. Sens Actuators B Chem 3:191–196
15. Fonstad CG, Linz A, Rediker RH (1969) Defect structure and electronic donor levels in stannic oxide crystals. J Electrochem Soc 116:1269–1273
16. Maffeïs TGG, Owen GT, Wilks SP, Malagù C, Martinelli G, Kennedy MK, Kruis FE (2004) Room temperature and elevated temperature characterisation of nanocrystalline SnO_2 surfaces by scanning tunnelling microscopy and spectroscopy. Int J Nanoscience 3:519–524
17. Guidi V, Butturi MA, Carotta MC, Cavicchi B, Ferroni M, Malagù C, Martinelli G, Vincenzi D, Sacerdoti M, Zen M (2002) Gas sensing through thick film technology. Sens Actuators B 84:72–77
18. Astolfi M, Zonta G, Gherardi S, Malagù C, Vincenzi D, Rispoli G (2023) A portable device for I-V and Arrhenius plots to characterize chemoresistive gas sensors: test on SnO_2-based sensors. Nanomaterials 13(18):2549
19. Comini E, Guidi V, Malagù C, Martinelli G, Pan Z, Sberveglieri G, Wang ZL (2004) Electrical properties of tin dioxide two-dimensional nanostructures. J Phys Chem B 108:1882–1887
20. Carotta MC, Benetti M, Ferrari E, Giberti A, Malagù C, Nagliati M, Vendemiati B, Martinelli G (2007) Basic interpretation of thick film gas sensors for atmospheric application. Sens Actuators B 126:672–677

21. Barsan N, Schweizer-Berberich M, Göpel W (1999) Fundamental aspects in the gas sensing mechanism of SnO_2-based sensors. Fresenius J Anal Chem 365:287–304
22. Zakrzewska K, Radecka M (2017) TiO_2-based nanomaterials for gas sensing—influence of anatase and rutile contributions. Nanoscale Res Lett 12:89
23. Setaro A, Lettieri S, Diamare D, Maddalena P, Malagù C, Carotta MC, Martinelli G (2008) Nanograined anatase titania-based optochemical gas detection. New J Phys 10:053030
24. Bonini N, Carotta MC, Chiorino A, Guidi V, Malagù C, Martinelli G, Paglialonga L, Sacerdoti M (2000) Doping of a nanostructured titania thick-film: structural and electrical investigations. Sens Actuators B 68:274–280
25. Carotta MC, Martinelli G, Crema L, Malagù C, Merli M, Ghiotti G, Traversa E (2001) Nanostructured thick film gas sensors for atmospheric pollutant monitoring: quantitative analysis on field tests. Sens Actuators B 76:336–342
26. Martinelli G, Carotta MC, Malagù C (2000) Physics and technology of thick film sensors and their applications for environmental gas monitoring. Electron Technol 33:40–44
27. Nunes D, Fortunato E, Martins R (2022) Flexible nanostructured TiO2-based gas and UV sensors: a review. Discover Mater 2:2
28. Gherardi S, Astolfi M, Gaiardo A, Malagù C, Rispoli G, Vincenzi D, Zonta G (2024) Investigating the temperature-dependent kinetics in humidity-resilient tin–titanium-based metal oxide gas sensors. Chemosensors 12(8):151
29. Guidi V, Boscarino D, Comini E, Faglia G, Ferroni M, Malagù C, Martinelli G, Rigato V, Sberveglieri G (2000) Preparation and characterisation of titanium-tungsten sensors. Sens Actuators B 65:264–266
30. Giberti A, Carotta MC, Malagù C, Aldao CM, Castro MS, Ponce MA, Parra R (2011) Permittivity measurements in nanostructured TiO_2 gas sensors. Phys Status Solidi A 208:118–122
31. Carotta MC, Gherardi S, Malagù C, Nagliati M, Vendemiati B, Martinelli G, Sacerdoti M, Lesci IG (2007) Comparison between titania thick films obtained through sol-gel and hydrothermal synthetic processes. Thin Solid Films 515:8339–8344
32. Carotta MC, Guidi V, Malagù C, Vendemiati B, Zanni A, Martinelli G, Sacerdoti M, Licoccia S, Di Vona ML, Traversa E (2005) Vanadium and tantalum-doped titanium oxide (TiTaV): a novel material for gas sensing. Sens Actuators B 108:89–96
33. Carotta MC, Ferroni M, Gherardi S, Guidi V, Malagù C, Martinelli G, Sacerdoti M, Di Vona ML, Licoccia S, Traversa E (2004) Thick-film gas sensors based on vanadium–titanium oxide powders prepared by sol-gel synthesis. J Eur Ceram Soc 24:1409–1413
34. Astolfi M, Rispoli G, Anania G, Artioli E, Nevoso V, Zonta G, Malagù C (2021) Tin, titanium, tantalum, vanadium and niobium oxide based sensors to detect colorectal cancer exhalations in blood samples. Molecules 26(2):46
35. Carotta MC, Gherardi S, Guidi V, Malagù C, Martinelli G, Vendemiati B, Sacerdoti M, Ghiotti G, Morandi S (2009) Electrical and spectroscopic properties of $Ti_{0.2}Sn_{0.8}O_2$ solid solution for gas sensing. Thin Solid Films 517:6176–6183
36. Schipani F, Aldao CM, Ponce MA (2012) Schottky barrier measurements through Arrhenius plots in gas sensors based on semiconductor films. AIP Adv 2:032138
37. Ponte R, Rauwel E, Rauwel P (2023) Tailoring SnO_2 defect states and structure: bottom-up approaches for energy applications. Materials 16:4339
38. Astolfi M, Rispoli G, Gherardi S, Zonta G, Malagù C (2023) Reproducibility and repeatability tests on $(SnTiNb)O_2$ sensors in detecting ppm-concentrations of CO and up to 40% of humidity: a statistical approach. Sensori 23(4):1983
39. Carotta MC, Gherardi S, Guidi V, Malagù C, Martinelli G, Vendemiati B, Sacerdoti M, Ghiotti G, Morandi S, Bismuto A, Maddalena P, Setaro A (2008) $(Ti, Sn)O_2$ binary solid solutions for gas sensing: spectroscopic, optical and transport properties. Sens Actuators B 130:38–45
40. Carotta MC, Fioravanti A, Gherardi S, Ghiotti G, Morandi S, Giberti A, Malagù C (2014) (Ti,Sn) solid solutions as functional materials for gas sensing. Sens Actuators B Chem 194:195–205
41. Giberti A, Malagù C, Guidi V (2012) WO_3 sensing properties enhanced by UV illumination: an evidence of surface effect. Sens Actuators B Chem 165:59–61
42. Wyckoff RWG (1964) Crystal structures, 2nd edn, vol 2. John Wiley & Sons

43. Staerz A, Somacescu S, Epifani M, Kida T, Weimar U, Barsan N (2020) WO_3-based gas sensors: identifying inherent qualities and understanding the sensing mechanism. ACS Sens 5:1624–1633
44. Blo M, Carotta MC, Galliera S, Gherardi S, Giberti A, Guidi V, Malagù C, Martinelli G, Sacerdoti M, Vendemiati B, Zanni A (2004) Synthesis of pure and loaded powders of WO_3 for NO_2 detection through thick film technology. Sens Actuators B 103:213–218
45. Ponzoni A, Comini E, Concina I, Ferroni M, Falasconi M, Gobbi E, Sberveglieri V, Sberveglieri G (2012) Nanostructured metal oxide gas sensors: applications at SENSOR lab. Sensors 12:17023–17045
46. Li X, Fu L, Karimi-Maleh H, Chen F, Zhao S (2024) Innovations in WO_3 gas sensors: nanostructure engineering and future perspectives. Heliyon 10:e27740
47. Kanda K, Maekawa T (2005) Development of a WO_3 thick-film-based sensor for VOC detection. Sens Actuators B Chem 108:97–101
48. Parellada-Monreal L, Gherardi S, Zonta G, Malagù C, Casotti D, Cruciani G, Guidi V, Martínez-Calderón M, Castro-Hurtado I, Gamarra D, Lozano J, Presmanes L, Mandayo GG (2020) WO_3 processed by direct laser interference patterning for NO_2 detection. Sens Actuators B Chem 305:127226
49. Malagù C, Carotta MC, Gherardi S, Guidi V, Vendemiati B, Martinelli G (2005) AC measurements and modeling of WO_3 thick film gas sensors. Sens Actuators B 108:70–74
50. Spagnoli E, Krik S, Fabbri B, Valt M, Ardit M, Gaiardo A, Vanzetti L, Della Ciana M, Cristino V, Vola G, Caramori S, Malagù C, Guidi V (2021) Development and characterization of WO_3 nanoflakes for selective ethanol sensing. Sens Actuators B Chem 347:130593
51. Malagù C, Carotta MC, Cervi A, Guidi V, Martinelli G (2007) Morphological differences affecting the dielectric response of MoO_3-WO_3 and WO_3. J Appl Phys 101:104310
52. Ghiotti G, Chiorino A, Prinetto F, Carotta MC, Malagù C (2001) Preparation and characterization of WO_x/SnO_2 nanosized powders for thick films gas sensors. Stud Surf Sci Catal 140:287–296
53. Chiorino A, Ghiotti G, Prinetto F, Carotta MC, Malagù C, Martinelli G (2001) Preparation and characterization of SnO_2 and WO_x-SnO_2 nanosized powders and thick films for gas sensing. Sens Actuators B 78:89–97
54. Jagadish C, Pearton SJ (2006) Zinc oxide bulk, thin films and nanostructures. Elsevier
55. Özgür Ü, Alivov YA, Liu C et al (2005) A comprehensive review of ZnO materials and devices. J Appl Phys 98:041301
56. Zonta G, Astolfi M, Casotti D et al (2020) Reproducibility tests with zinc oxide thick film sensors. Ceram Int 46:6847–6855
57. Zappa D, Bertuna A, Comini E, Kaur N, Poli N, Sberveglieri V, Sberveglieri G (2017) Metal oxide nanostructures: preparation, characterization and functional applications as chemical sensors. Beilstein J Nanotechnol 8:1205–1217
58. Comini E, Sberveglieri G (2010) Metal oxide nanowires as chemical sensors. Mater Today 13:36–44
59. Baruwati B, Kumar DK, Manorama SV (2006) Hydrothermal synthesis of highly crystalline ZnO nanoparticles. Sens Actuators B Chem 119:676–682
60. Agarwal S, Rai P, Gatell EN et al (2019) Gas sensing properties of ZnO nanostructures synthesized by hydrothermal method. Sens Actuators B Chem 292:24–31
61. Aneesh PM, Vanaja KA, Jayaraj MK (2007) Synthesis of ZnO nanoparticles by hydrothermal method. Proc SPIE 6639:66390J
62. Komorizono AA, Tagliaferro JC, Mastelaro VR (2025) Ozone sensing capabilities of porous ZnO nanostructures. Mater Lett 388:138332
63. Christaki E, Vasilaki E, Gagaoudakis E et al (2022) Room temperature optical detection of ozone using ZnO nanohybrids. Sens Actuators B Chem 359:131614
64. Bernardini S, Benchekroun MH, Fiorido T et al (2017) Ozone sensors using ZnO nanocrystals. Proceedings 1:423
65. Park S, An S, Mun Y, Lee C (2013) UV-enhanced NO_2 sensing of SnO_2/ZnO nanowires. ACS Appl Mater Interfaces 5:4285–4292

66. Geng Q, He Z, Xun C, Dai W, Wang X (2013) Gas sensing of ZnO under visible light. Sens Actuators B Chem 188:293–297
67. Gaiardo A, Fabbri B, Giberti A et al (2016) ZnO and Au/ZnO thin films: room-temperature chemoresistive properties. Sens Actuators B Chem 237:1085–1094
68. Fabbri B, Gaiardo A, Giberti A et al (2016) Chemoresistive properties of photo-activated ZnO films. Sens Actuators B Chem 222:1251–1256
69. Guidi V, Fabbri B, Gaiardo A, Gherardi S, Giberti A, Malagù C, Zonta G, Bellutti P (2015) Metal sulfides as a new class of sensing materials. Proce Eng 120:138–141. ISSN 1877-7058
70. Gaiardo A, Fabbri B, Guidi V, Bellutti P, Giberti A, Gherardi S, Vanzetti L, Malagù C, Zonta G (2016) Metal sulfides as sensing materials for chemoresistive gas sensors. Sensors 16:296
71. Giberti A, Casotti D, Cruciani G, Fabbri B, Gaiardo A, Guidi V, Malagù C, Zonta G, Gherardi S (2015) Electrical conductivity of CdS films for gas sensing: selectivity properties to alcoholic chains. Sens Actuators B Chem 207:504–510. https://doi.org/10.1016/j.snb.2014.10.116
72. Giberti A, Fabbri B, Gaiardo A, Guidi V, Malagù C (2014) Resonant photoactivation of cadmium sulfide and its effect on the surface chemical activity. Appl Phys Lett 104:222102
73. Giberti A, Gaiardo A, Fabbri B, Gherardi S, Guidi V, Malagù C, Bellutti P, Zonta G, Casotti D, Cruciani G (2016) Tin (IV) sulfide nanorods as a new gas sensing material. Sens Actuators B Chem 223:827–833
74. Rodriguez J, McCarthy M (2022) Thermistor-based temperature sensing system: design challenges and circuit configuration. Analog dialogue, analog devices
75. Ortega PP, Aparecido R, Amoresi C, Daldin M, Merízio LG, Ramirez MA, Aldao CM, Malagù C, Ponce MA, Longo E, Simões AZ (2024) Insights into morphology and structural defects of Eu-doped ceria nanostructures for optoelectronic applications in red-emitting devices. ACS Appl Nano Mater 7(11):12466–12479. https://doi.org/10.1021/acsanm.4c00875
76. Ortega PP, Gherardi S, Spagnoli E, Fabbri B, Astolfi M, Zonta G, Landini N, Malagù C, Aldao CM, Ponce MA, Simoes AZ, Longo E (2024) Nanostructured Eu-doped ceria for humidity sensing: a morphological perspective. Appl Surf Sci 661:160047
77. Vincenzi D, Butturi MA, Stefancich M, Malagù C, Guidi V, Carotta MC, Martinelli G, Guarnieri V, Brida S, Margesin B, Giacomozzi F, Zen M, Vasiliev AA, Pisliakov AV (2001) Low-power thick-film gas sensor obtained by a combination of screen printing and micromachining techniques. Thin Solid Films 391:288–292
78. Montoncello F, Carotta MC, Cavicchi B, Ferroni M, Giberti A, Guidi V, Malagù C, Martinelli G, Meinardi F (2003) Near-infrared photoluminescence in titania: an evidence for phonon-replica effect. J Appl Phys 94:1501–1505
79. Carotta MC, Ferroni M, Guidi V, Martinelli G (1999) Gas sensing properties of nanostructured metal oxides. Adv Mater 11:943–946
80. Guidi V, Blo M, Butturi MA, Carotta MC, Galliera S, Giberti A, Malagù C, Martinelli G, Piga M, Sacerdoti M, Vendemiati B (2004) Aqueous and alcoholic syntheses of tungsten trioxide powders for NO_2 detection. Sens Actuators B 100:277–282
81. Gaiardo A, Fabbri B, Giberti A, Valt M, Gherardi S, Guidi V, Malagù C, Bellutti P, Pepponi G, Casotti D, Cruciani G, Zonta G, Landini N, Barozzi M, Morandi S, Vanzetti L, Canteri R, Della Ciana M, Migliori A, Demenev E (2020) Tunable formation of nanostructured SiC/SiOC core-shell for selective detection of SO_2. Sens Actuators B Chem 305:127485
82. Giberti A, Carotta MC, Fabbri B, Gherardi S, Guidi V, Malagù C (2012) High-sensitivity detection of acetaldehyde. Sens Actuators B Chem 174:402–405
83. Morandi S, Prinetto F, Di Martino M, Ghiotti G, Lorret O, Tichit D, Malagù C, Vendemiati B, Carotta MC (2006) Synthesis and characterisation of gas sensor materials obtained from Pt/Zn/Al layered double hydroxides. Sens Actuators B 118:215–220
84. Carotta MC, Cervi A, Giberti A, Guidi V, Malagù C, Martinelli G, Puzzovio D (2009) Ethanol interference in light alkane sensing by metal-oxide solid solutions. Sens Actuators B Chem 136:405–409
85. Fabbri B, Bonoldi L, Guidi V, Cruciani G, Casotti D, Malagù C, Bellussi G, Millini R, Montanari L, Carati A, Rizzo C, Montanari E, Zanardi S (2017) Crystalline microporous organosilicates with reversed functionalities of organic and inorganic components for room-temperature gas sensing. ACS Appl Mater Interfaces 9:24812–24820

86. Wagner T, Kohl CD, Malagù C, Donato N, Latino M, Neri G, Tiemann M (2013) UV light-enhanced NO_2 sensing by mesoporous In_2O_3: Interpretation of results by a new sensing model. Sens Actuators B Chem 187:488–494
87. Carotta MC, Cervi A, Gherardi S, Guidi V, Malagù C, Martinelli G, Vendemiati B, Sacerdoti M, Ghiotti G, Morandi S, Lettieri S, Maddalena P, Setaro A (2009) $(Ti, Sn)O_2$ solid solutions for gas sensing: a systematic approach by different techniques for different calcination temperature and molar composition. Sens Actuators B Chem 139:329–339
88. Carotta MC, Cervi A, Giberti A, Guidi V, Malagù C, Martinelli G, Puzzovio D (2008) Metal-oxide solid solutions for light alkane sensing. Sens Actuators B Chem 133:516–520
89. Rampino S, Pattini F, Malagù C, Pozzetti L, Stefancich M, Bronzoni M (2014) Application of a substrate bias to control the droplet density on $Cu(In,Ga)Se_2$ thin films grown by pulsed electron deposition. Thin Solid Films 562:307–313
90. Spagnoli E, Gaiardo A, Fabbri B, Valt M, Krik S, Ardit M, Cruciani G, Della Ciana M, Vanzetti L, Vola G, Gherardi S, Bellutti P, Malagù C, Guidi V (2022) Design of a metal-oxide solid solution for sub-ppm H_2 detection. ACS Sens 7:573–583

Chapter 9
Temperature and Humidity Sensors

It is not the strongest of the species that survive, nor the most intelligent,
but the one most responsive to change.
Charles Darwin

Temperature and humidity sensors play a fundamental role in the context of chemoresistive gas sensing, as the response of gas-sensitive materials is intrinsically affected by external environmental conditions. In this Chapter, attention is devoted to temperature and humidity sensors, which are indispensable for both the accurate interpretation and stabilization of gas sensor outputs. Variations in temperature influence the adsorption-desorption dynamics of gas molecules, modify the carrier concentration within semiconducting sensing layers, and alter reaction kinetics at the sensor surface, thereby shifting baseline resistance and sensitivity. Similarly, humidity exerts a complex effect on gas sensors: water molecules can compete with target analytes for active surface sites, form hydroxyl species that modify surface charge, and even alter the microstructure or conduction pathways within porous sensing films. Therefore, integrating precise thermal and humidity monitoring is essential for the reliable operation, calibration, and compensation of chemoresistive devices, especially under real-world conditions where these environmental parameters fluctuate continuously. This Chapter outlines the working principles of the main thermal and humidity sensors, emphasizing their synergistic importance in advanced gas-sensing systems and environmental monitoring platforms. This Chapter is grounded in a comprehensive and integrated review of the relevant literature cited here, and extends it through the authors' original contributions [1–16].

9.1 Heat Transmission

The transmission of heat, or heat transfer, represents the physical process through which thermal energy is exchanged between bodies or within a body due to a temperature difference. Fundamentally, there are three mechanisms of heat transmission: conduction, convection, and radiation. Each of these processes is governed by distinct

© The Author(s), under exclusive license to Springer Nature Switzerland AG 2026

C. Malagù and G. Zonta, *Gas Sensors*, https://doi.org/10.1007/978-3-032-21614-4_9

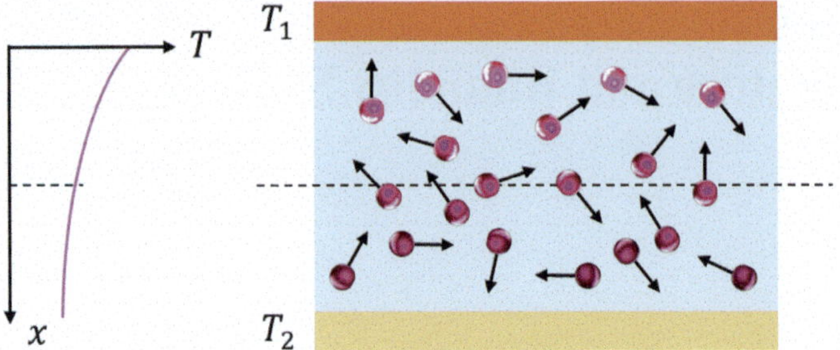

Fig. 9.1 Schematization of two surfaces (at temperature T_1 and T_2, respectively, with $T_1 > T_2$) in contact with a gas. The particles are indicated as pink balls with their luminosity increasing with temperature. The particles in the hotter region are characterized by higher kinetic energy and collide with less energetic particles in the cooler region, transferring part of their energy until thermal equilibrium is reached

physical principles and plays a crucial role in determining the thermal behavior of materials and systems.

Conduction occurs when heat is transferred through direct physical contact between particles of matter. This mechanism, schematized for a gas in Fig. 9.1, relies on the microscopic interaction between particles in regions of different temperatures: those in the hotter region, characterized by higher kinetic energy, collide with less energetic particles in the cooler region, transferring part of their energy until thermal equilibrium is reached. In gases, this process is known as thermal diffusion, and it results from the random motion and collisions of energetic molecules across a temperature gradient.

In **solids**, however, the situation is quite different because atoms are not free to move as in a gas, being bound to fixed positions in a crystal lattice. Nevertheless, heat can still propagate through the material due to the atoms' vibrational energy. As the temperature rises, atoms vibrate more intensely around their equilibrium positions, and these vibrations propagate through the lattice as phonons, quanta of lattice vibrations; this mechanism is known as phonon conduction. In metals, an additional contribution arises from the free electrons, which often provide the dominant contribution to thermal conductivity. The quantitative description of conduction is provided by **Fourier's Law**:

$$\vec{H} = -k\nabla T,$$

where $\vec{H}$ is the heat flux vector (measured in W/m^2), k is the thermal conductivity of the material (in W/(m K^{-1}), and ∇T is the temperature gradient. The negative sign indicates that heat flows spontaneously from hotter to cooler regions. In one dimension, this becomes:

$$H = \frac{dQ}{dt} = -kA\frac{dT}{dx},$$

where A is the cross-sectional area through which heat flows. The value of k depends strongly on the material: for instance, aluminum exhibits a high conductivity ($k \approx 205$ W/m K^{-1}), while air, being a poor conductor, shows a much lower value ($k \approx 0.024$ W/m K^{-1}). In Table 9.1 a comparison between the thermal conductivity of different materials.

Convection is another mechanism of heat transfer that involves a fluid medium (a gas or liquid) which acts as an intermediate agent carrying thermal energy from one region to another. Unlike conduction, where heat transfer occurs only through molecular collisions, convection involves both microscopic diffusion and macroscopic mass motion. This latter mechanism, corresponding to the bulk movement of fluid parcels, typically dominates and greatly enhances the heat transfer process. Convection can occur in two distinct forms:

- *natural* (or free) convection, when the movement of the fluid arises spontaneously from density gradients induced by temperature differences (i.e., warmer, less dense fluid rises while cooler, denser fluid sinks);
- *forced* convection, when an external agent (such as a fan or pump) induces the flow, thereby increasing the rate of heat transfer.

In both cases, the rate of heat transfer can be expressed as proportional to the temperature difference between surface and fluid and the exposed surface area, through the convective heat transfer coefficient (α), measured in W m^{-2} K^{-1}:

$$\frac{dQ}{dt} = \alpha A\left(T_s - T_f\right),$$

where T_s and T_f are the temperatures of the solid surface and the fluid, respectively. Convection plays an essential role in many natural and technological systems, from atmospheric circulation to cooling systems in electronic and chemical

Table 9.1 Comparative table between the thermal conductivity of different materials

Material type	Examples	Thermal conductivity (W/m K^{-1})
Vacuum systems	Vacuum-insulated panels (VIP), cryogenic insulation	0.001–0.01
Multi-layer insulation (MLI, vacuum)	Thermal blankets, spacecraft insulation	0.0001–0.005
Powders, glass fibers	Aerogels, fiberglass, mineral wool	0.02–0.06
Polymers	PVC, PTFE, polyethylene, epoxy resins	0.15–0.4
Glass/ceramic foams	Porous SiO_2 foams	0.03–0.1
Glass, low- k ceramics	Fused silica, glass, porcelain	1–2
Dense ceramics	Alumina (Al_2O_3), zirconia	20–35
Crystalline solids (metals)	Cu, Al, Ag, Fe, W	50–400

processes. **Thermal radiation** represents the last mechanism of heat transfer, and it is intrinsically different from conduction and convection, as it does not require any material medium. Instead, heat is emitted in the form of electromagnetic waves, or photons, which carry energy away from the emitting surface. Every body with a temperature above absolute zero ($T > 0$ K) emits thermal radiation due to the random motion of its charged particles. The total emissive power of a surface, i.e., the radiant energy emitted per unit area, is governed by the **Stefan–Boltzmann law**:

$$E_b = \sigma_0 T_s^4,$$

where E_b is the emissive power of an ideal black body, T_s is the absolute temperature (in K), and $\sigma_0 = 5.67 \times 10^{-8}$ W m^{-2} K^{-4} is the Stefan–Boltzmann constant. A black body represents an idealized surface that absorbs and emits the maximum possible radiation at a given temperature. Real surfaces emit less radiation, and their efficiency is characterized by the *emissivity* (ε), a dimensionless property ranging from 0 (perfect reflector) to 1 (ideal emitter). The actual emitted power is thus:

$$E = \varepsilon \sigma_0 T_s^4.$$

Radiation is unique among the three mechanisms since it can transfer energy through a vacuum, making it the dominant heat transfer mode in space. Furthermore, the spectral distribution of radiation depends on temperature, as described by Planck's law, while the Wien's displacement law predicts that the peak emission wavelength shifts toward shorter wavelengths as the temperature increases, explaining why heated bodies change color from red to white as they get hotter.

In conclusion, heat transmission arises from a complex interplay of microscopic and macroscopic mechanisms: conduction via particle or phonon interactions, convection through mass motion in fluids, and radiation through electromagnetic emission. The understanding of heat transmission mechanisms is of crucial importance in gas sensor measurements, as thermal exchanges within the measurement chamber can significantly influence the effective temperature of the sensing layer. In experimental practice, it has been demonstrated by Giberti et al. [7] that the film conductance is not solely dependent on the nominal operating temperature of the sensor but rather on a complex interplay between the temperature of the film itself and the temperature of the air nearby the film. In Fig. 9.2 a sensor schematized with the main components affecting the sensing film conductance.

This is because heat transfer within the sensing system occurs simultaneously through conduction, convection, and radiation, each contributing differently depending on the materials, geometry, and boundary conditions of the setup. Conduction governs the heat flow across the sensor and its metallic support, convection mediates energy exchange with the ambient air, and radiation becomes relevant at higher temperatures or in low-pressure environments. To accurately model the thermal behavior and ensure reliable gas sensing performance, it is therefore necessary to consider all these mechanisms by defining three coupled thermal balance equations that describe the dynamic equilibrium between the sensor layer (T_{sensor}),

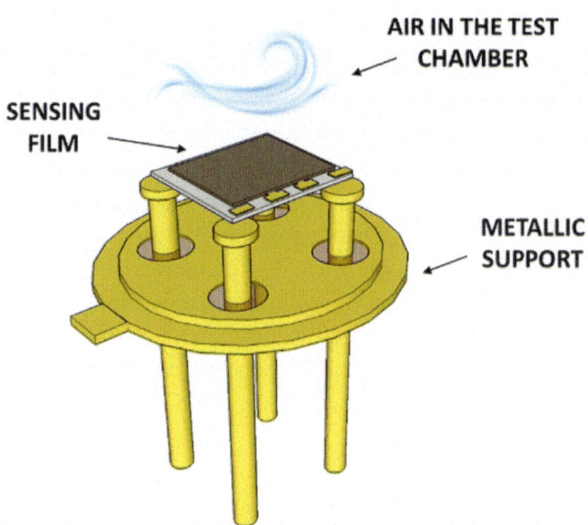

Fig. 9.2 Schematization of a chemoresistive sensor with the main components affecting the sensing film conductance

the metallic support (T_{support}), and the test chamber (T_{chamber}). A proper understanding and control of these parameters allow one to decouple the intrinsic sensor response from parasitic thermal effects, leading to more reproducible and physically meaningful measurements.

9.2 Temperature Sensors

All thermal sensors, irrespective of their underlying measurement principle, share several intrinsic characteristics that influence their overall performance and accuracy. A fundamental common aspect to all of them (except non-contact sensors based on thermal radiation) is the necessity of physical contact between the sensor and the body whose temperature (T_X) is to be measured. This contact inevitably introduces several sources of error that compromise the precision of the temperature reading. Firstly, the sensor does not immediately attain the temperature of the body due to the finite time required to reach thermal equilibrium. Secondly, the act of contact itself modifies the body's temperature, as the sensor possesses a non-zero thermal capacity (C_S) that allows it to absorb or release heat, thereby slightly altering the true temperature of the body. Moreover, the existence of a finite thermal resistance between the sensor and the body, denoted as $R_{XS}(T)$, delays the establishment of thermal equilibrium and contributes to transient temperature deviations. In addition, the finite thermal resistance between the sensor and the surrounding environment, $R_{SE}(T)$, permits heat exchange between the environment and the sensor, which can further disturb the temperature of the body under observation. Typically:

$$R_{XS}(T) \ll R_{SE}(T),$$

ensuring that the heat exchange between the sensor and the environment is minimal compared to that between the sensor and the body. Furthermore, the functioning of thermal sensors may itself generate heat, which flows into the body under measurement. In thermal modeling, these heat flows can be represented as ideal current generators, and each physical body (i.e., sensor, measured object, environment) can be associated with a network node whose voltage corresponds to its temperature. The node voltages, representing temperatures, are measured with respect to an auxiliary node (ground). Each node also possesses a capacitance toward ground, equivalent to the thermal capacity of the respective body. The thermal capacity of an object quantifies the amount of heat that can be stored as kinetic energy within its atomic or molecular structure. It is expressed as:

$$C = c \cdot m,$$

where c is the specific heat (J kg^{-1} K^{-1}) and m is the mass of the body. Thus, the thermal behavior of a system can be effectively analyzed by an equivalent electrical circuit, where heat flow corresponds to electric current, temperature corresponds to voltage, thermal resistances model heat transfer limitations, and thermal capacities describe heat storage. This analogy provides a powerful framework for understanding and optimizing sensor dynamics and their interaction with the measured system. In Fig. 9.3 a scheme representing heath flows schematized as ideal current generators in an electronic circuit.

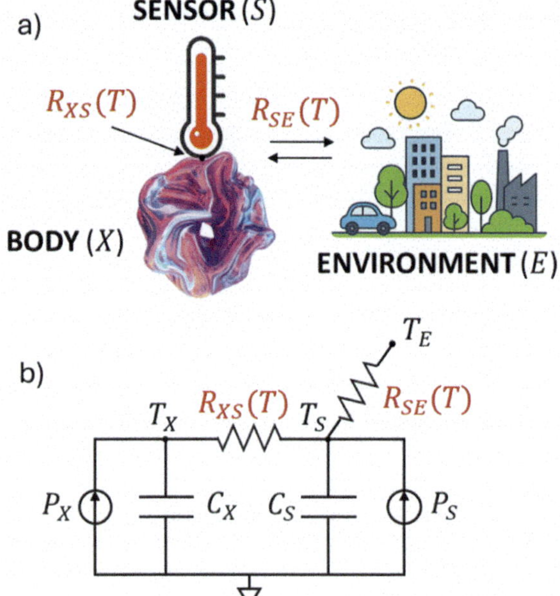

Fig. 9.3 **a** Scheme of a thermal sensor (S) in contact with a body (X) inside an external environment (E). $R_{XS}(T)$ and $R_{SE}(T)$ are the thermal resistances between the body and the sensor and between the sensor and the environment, respectively; **b** heat flows between the body and the sensor schematized ideal current generators (P_X and P_S) in an electronic circuit, where C_X and C_S are their thermal capacities, T_X and T_S their temperatures and T_E the temperature of the environment

Over the years, several types of temperature sensors have been developed, each based on different physical principles and optimized for specific operating conditions, accuracy requirements, and environmental constraints. The most widely used temperature sensors include:

- thermocouples;
- resistance temperature detectors (RTDs);
- thermistors;
- semiconductor temperature sensors;
- p-n junction sensors;
- pyrometers.

Thermocouples are temperature-sensing device composed of a couple of electric conductors made from different materials that exhibit distinct thermoelectric properties. These two wires are joined at one end, forming the so-called hot junction (or measuring junction), while their opposite ends, referred to as the cold junction (or reference junction), are maintained at a different temperature. The temperature difference between the two junctions generates an electromotive force *emf*, according to the Seebeck effect, which can be measured by an ideal voltmeter or an amplifier with high input impedance. This *emf*, proportional to the temperature difference, forms the basis of thermoelectric temperature measurement. When one end of a conductor is heated, the electrons in the hotter region acquire greater kinetic energy compared to those in the colder part. As a result, energetic electrons tend to diffuse from the hot junction toward the colder region, where the available energy levels are lower. This diffusion leads to a charge imbalance: negative charges accumulate near the cold side, while a deficiency of electrons develops near the hot side. The process continues until the electrostatic potential created by this separation of charges counterbalances further diffusion. Under these conditions of dynamic equilibrium, an internal electric field is established within the conductor.In a thermocouple composed of two dissimilar conductors, the magnitude of this thermoelectric potential differs for each material due to their distinct electronic structures. Consequently, when the two conductors are connected to form a closed loop, a net electromotive force appears between the open ends, proportional to the temperature difference between the two junctions. The resulting thermoelectric current can thus be used to infer the temperature of the measuring junction relative to the reference junction. In Fig. 9.4, a scheme of a thermocouple composed of two wires of different materials connecting two objects at different temperatures (T) and ($T + \Delta T$) and an ideal voltmeter that measures a potential difference ΔV.

For accurate measurements, it is essential that both reference junctions are maintained at the same and known temperature, typically achieved through cold-junction compensation, as in Fig. 9.5.

The wires forming the thermocouple, denoted as A and B, should possess a large variation in *emf* with respect to temperature, ensuring high sensitivity. The conductor C serves merely as an electrical connection to the amplifier (G) or to the voltmeter that reads the generated output voltage, without affecting the thermoelectric behavior of

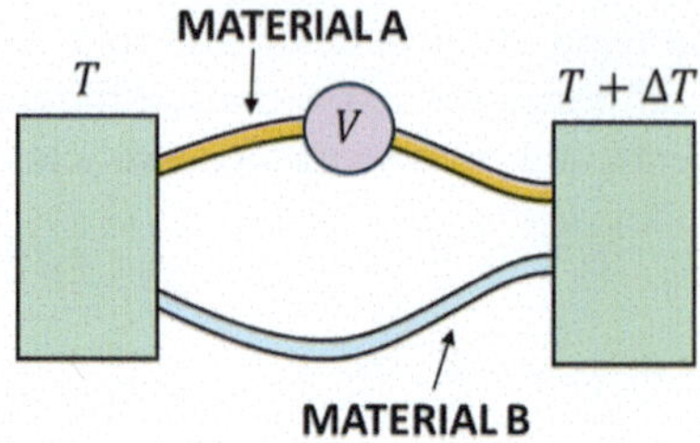

Fig. 9.4 Scheme of the thermocouple working principle. Two wires of different materials connect two objects at different temperatures (T) and ($T + \Delta T$) and an ideal voltmeter that measures a potential difference ΔV, which is correlated with the two bodies ΔT

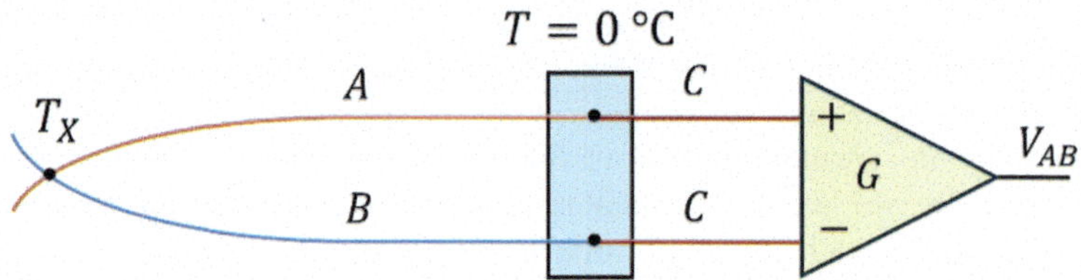

Fig. 9.5 Scheme of a thermocouple where A and B are the two different material wires, connected with conductors C to the amplifier G and to the voltmeter that reads the output voltage V_{AB}. The cold junction is maintained at $T = 0$ °C (e.g., by a mixture of ice and water), while the hot junction measures the body temperature T_X

the sensing pair. The relation between the output voltage and the temperature difference is nonlinear and depends on the thermocouple type but it can be approximated as a polynomial equation like:

$$\Delta T = \sum_{i=0}^{N} a_i V^i,$$

where the a_i coefficient vary depending on the materials employed. Commercial thermocouples exhibit characteristic voltage-temperature relationships depending on the materials used. Typical thermocouple types include: K (Chromel-Alumel), J (Iron-Constantan), T (Copper-Constantan), and S (Platinum-Rhodium), each designed for specific temperature ranges and environments. The generated voltage typically spans from a few microvolts per degree Celsius (μV/ °C) for noble-metal thermocouples to several tens of μV/ °C for base-metal ones.**Resistance Temperature Detectors (RTDs)** are precision temperature sensors that operate on the principle of the variation of electrical resistance with temperature in metallic conductors. As temperature increases, the amplitude of lattice vibrations rises, leading to more frequent electron scattering events and, consequently, to an increase in resistivity. This dependence between resistance and temperature is highly stable, predictable, and, over a wide range, nearly linear. An RTD typically consists of a metallic wire or a thin metallic film, most often made of platinum, deposited or wound on an insulating substrate. Platinum is preferred due to its chemical inertness, excellent stability, and reproducible resistance-temperature relationship, which remain substantially unchanged even after long-term operation or repeated thermal cycling. The resistivity of platinum at room temperature is $\rho = 10.6 \times 10^{-8}$ Ωm, approximately six times higher than that of copper ($\rho_{Cu} = 1.7 \times 10^{-8}$ Ωm). This relatively high resistivity allows for achieving significant resistance values even with short or thin sensing elements, improving compactness and sensitivity. The **temperature coefficient of resistance (TCR)**, defined as:

$$\alpha = \frac{1}{R}\frac{dR}{dT},$$

for platinum assumes a typical value of $\alpha = 3.85 \times 10^{-3}$ K^{-1} at 0 °C, slightly decreasing approximately to $\alpha = 2.90 \times 10^{-3}$ K^{-1} at 8000 °C. This modest variation ensures high linearity and predictability of the sensor's response. The most accurate and widely adopted empirical expression describing the relationship between resistance and temperature in platinum RTDs is given by the **Callendar–Van Dusen equation**:

$$R_T = R_0\left[1 + AT + BT^2 + C(T - 100)T^3\right],$$

where R_T is the resistance at temperature T (°C), R_0 is the resistance at 0 °C, and A, B, C are empirical coefficients determined by calibration. For temperatures $T > 0$ °C, the cubic correction term becomes negligible, and the coefficient C is set to zero, reducing the expression to a quadratic form:

$$R_T = R_0\left[1 + AT + BT^2\right].$$

Typical values of the Callendar-Van Dusen coefficients for platinum RTDs are:

$$A = 3.9083 \times 10^{-3} {}^{\circ}\mathrm{C}^{-1},$$

$$B = -5.775 \times 10^{-7} {}^{\circ}\mathrm{C}^{-2},$$

$$C = -4.183 \times 10^{-12} {}^{\circ}\mathrm{C}^{-4}.$$

This formulation provides an accurate fit of the resistance-temperature curve across a wide range of temperatures, ensuring compatibility with standardized calibration scales such as ITS-90 and IEC 60751. Regarding their performance and characteristics, RTDs offer exceptional accuracy, repeatability, and long-term stability, outperforming thermocouples in precision temperature measurement within moderate temperature ranges. Standard industrial RTDs, such as Pt100 (100 Ω at $T = 0$ °C) or Pt1000 (1000 Ω at $T = 0$ °C), are widely adopted and feature well-defined calibration curves traceable to international standards. Typical operating ranges extend from −200 to 850 °C, depending on construction type and encapsulation. Wire-wound RTDs provide superior accuracy and are preferred for laboratory and calibration applications, while thin-film RTDs offer robustness, faster response times, and better integration into compact electronic systems. RTDs combine several notable advantages:

- high accuracy (up to ±0.1 °C when calibrated);
- high reproducibility;
- excellent linearity in the 0–600 °C range;
- long-term stability and minimal drift;
- compatibility with a wide range of signal-conditioning electronics.

However, they also exhibit certain limitations: RTDs are more expensive and mechanically delicate compared to thermocouples, and their response time can be slower due to the higher thermal mass of the sensing element. Additionally, since RTDs require a small excitation current for resistance measurement, self-heating effects must be minimized to avoid measurement bias. **Thermistors** are temperature-dependent resistors whose electrical resistance varies strongly and non-linearly with temperature. The term itself originates from the combination of "thermal" and "resistor", indicating a material whose resistive properties are influenced by thermal excitation. Unlike RTDs, which exhibit an almost linear relationship between resistance and temperature, thermistors are characterized by an exponential dependence that makes them extremely sensitive to even small temperature variations. Thermistors are usually composed of semiconducting metal oxides (e.g., manganese, nickel, cobalt, copper, iron oxides) often mixed with additional elements to tailor the desired resistance and thermal characteristics. The manufacturing process typically involves the insertion of two metallic electrodes into the ceramic body before sintering, or alternatively, the deposition of a serigraphic film of thermistor material onto an insulating substrate such as alumina. This latter technology allows for the fabrication of compact, planar sensors compatible with modern electronic integration. Depending on the sign of their TCR, thermistors are classified into two main categories:

- *NTC (Negative Temperature Coefficient)* thermistor, which exhibits a decrease in resistance with increasing temperature and is primarily employed as a temperature sensor due to its high sensitivity and reproducibility;
- *PTC (Positive Temperature Coefficient)* thermistors, in which resistance increases with temperature; they are often based on doped ceramic materials and used in self-heating devices such as current limiters, thermostats, or overcurrent protection circuits.

The electrical behavior of an NTC thermistor can be described by the empirical thermistor law:

$$R(T) = R(T_0)\, e^{B\left(\frac{1}{T} - \frac{1}{T_0}\right)},$$

where $R(T)$ is the resistance at temperature T (expressed in K), $R(T_0)$ is the resistance at a reference temperature T_0 (typically $T_0 = 25\ °C = 298.15$ K), and B is the material constant (or characteristic temperature), with values generally between 2000 and 5000 K. From this relation, the TCR can be derived as:

$$\alpha = \frac{1}{R}\frac{dR}{dT} = -\frac{B}{T^2},$$

which shows a strong and non-linear dependence on temperature. At ambient temperature, the TCR typically reaches values around $\alpha = 4 \times 10^{-2}\ K^{-1}$, which is about one order of magnitude higher than that of platinum RTDs. This high sensitivity allows thermistors to detect even minute thermal variations. For example, a thermistor with $R_0 = 5000\ \Omega$ and a sensitivity of 4%/°C exhibits a resistance variation of about $\Delta R = 200\ \Omega$ for a temperature change of just $\Delta T = 1$ °C, a variation that is easily measurable even with simple readout electronics. Thermistors are thus highly sensitive, compact, and fast-responding temperature sensors, capable of providing reliable data across limited temperature intervals. However, their response is strongly non-linear, restricting their use over broad temperature ranges unless compensated electronically or digitally. Typical resistance values span from a few hundred Ω to several hundred kΩ at 25 °C, with operating temperature ranges generally between –50 and 150 °C, extending up to 300 °C for special high-temperature models. Because of their low cost, versatility, and ease of integration, thermistors are widely used in industrial process control, environmental sensing, and consumer electronics. They are key components in automotive systems, household appliances, and medical devices, as well as in temperature compensation circuits and battery management systems. While thermocouples are preferred for very wide temperature intervals and RTDs for precision measurements, thermistors represent the optimal solution for applications requiring high sensitivity, compact size, and economic efficiency within moderate temperature ranges. In Fig. 9.6 a graph of the resistance-vs-temperature curve of NTC, PTC detectors and RTD detectors.

p-n junction temperature sensors are based on the intrinsic temperature dependence of the electrical properties of a semiconductor junction formed between p- and n-type materials. When a p-n junction (such as that of a diode or a transistor base-emitter junction) is polarized under a constant current I_C, a voltage drop V is established across the junction that depends strongly on temperature. As the temperature increases, the forward voltage of the junction decreases due to enhanced carrier mobility and the reduction of the barrier potential within the depletion region. This voltage-temperature relationship is generally non-linear, but within certain operating intervals it can be considered approximately linear, which simplifies its practical use as a temperature sensor. The behavior can be qualitatively described by the diode equation:

$$I = I_S\left(e^{\frac{qV}{kT}} - 1\right),$$

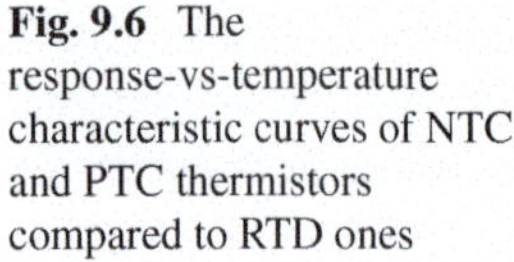
Fig. 9.6 The response-vs-temperature characteristic curves of NTC and PTC thermistors compared to RTD ones

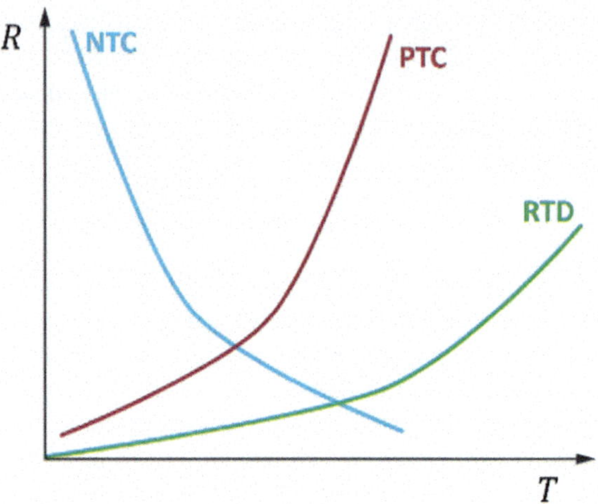

where I_S is the saturation current, q the electron charge, k the Boltzmann constant, T the absolute temperature, and V the voltage across the junction. By keeping the current I_C as a constant, the voltage V becomes inversely dependent on temperature according to:

$$V(T) \approx \frac{kT}{q} \ln\left(\frac{I_C}{I_S}\right).$$

Since I_S itself increases exponentially with temperature, the overall voltage decreases almost linearly as temperature rises. The typical sensitivity of such a junction is approximately −2 mV/°C, which means that for every degree Celsius increase in temperature, the forward voltage drops by about 2 mV. This characteristic makes p-n junctions valuable for temperature sensing applications, particularly in integrated circuits, where they can be easily fabricated alongside other electronic components with minimal additional cost. Their compactness and compatibility with semiconductor technology have led to their widespread use in on-chip thermal monitoring, temperature compensation circuits, and thermal management of electronic devices. One of the main advantages of p-n junction sensors is their operability at extremely low temperatures. Indeed, they can function efficiently down to cryogenic conditions, reaching temperatures as low as T~1.4 K. At such low temperatures, the slope of the voltage-temperature curve increases, leading to a marked enhancement in sensitivity. At $T < 40$ K, this effect becomes particularly evident, making these sensors well-suited for low-temperature physics experiments, cryogenic systems, and space applications, where precise temperature control at very low T is essential. In Fig. 9.7 a typical voltage-vs-temperature curve of a p-n junction sensor.

In summary, p-n junction temperature sensors exploit the temperature dependence of the forward voltage in a semiconductor diode under constant current bias. Despite their non-linear response, they offer high reproducibility, good sensitivity,

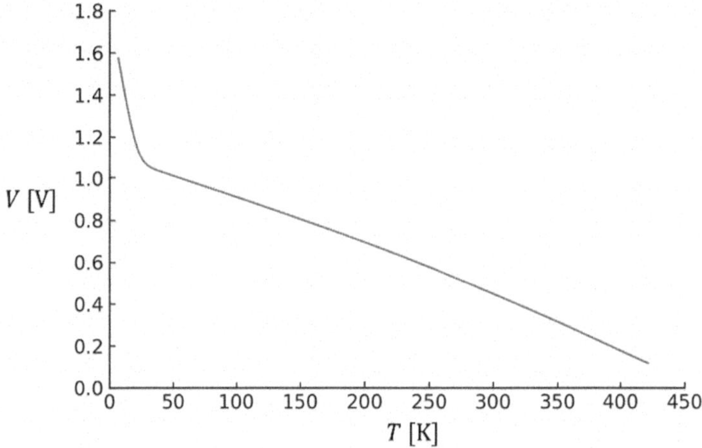

Fig. 9.7 Voltage-vs-curve trend of a typical p-n junction temperature sensor. It appears evident that V decreases with increasing T and that the dependence is non-linear, but below $T = 40$ K it can be observed an increase in slope (sensitivity)

and excellent integration capability, making them a practical and efficient choice for temperature measurements over a wide range from ambient conditions down to cryogenic levels.

Pyrometers are non-contact temperature sensors that operate based on the measurement of thermal radiation emitted by a body. They are, in essence, electrical infrared radiation transducers, exploiting the physical principle that any object at a temperature above absolute zero emits electromagnetic radiation whose spectral distribution and intensity depend on its temperature, as described by Planck's law. Unlike contact sensors, pyrometers determine temperature without the need for physical contact with the target, making them especially valuable in high-temperature or rapidly changing environments where intrusive measurements are impractical or impossible. As seen in Sect. 9.1, in an ideal case, the total radiant power emitted per unit surface area by a perfect blackbody is given by the Stefan-Boltzmann law, obtained by integrating Planck's law over all wavelengths, while for a generic body at a temperature T, it depends on the emissivity ε ($E = \varepsilon\sigma_0 T_s^4$). In practical implementations, pyrometers do not receive radiation across all wavelengths, as the optical components (i.e., lenses and mirrors) used to focus the thermal radiation onto the sensing element act as spectral filters. These materials are typically opaque to longwave and shortwave infrared radiation but transparent in the visible and near-infrared regions. Consequently, the sensor does not integrate the entire spectrum of Planck's radiation law, but only a selected portion of it, effectively measuring a narrowband radiative flux that still retains a monotonic relationship with temperature. Because the radiative power varies approximately as T^4, the response of the pyrometer is intrinsically non-linear. However, this non-linearity can be managed through calibration procedures, allowing the device to deliver accurate temperature readings across a defined range. The output signal (an electrical voltage or current proportional to the detected radiation intensity) is then converted into a temperature value through appropriate algorithms or analog circuitry. Pyrometers offer several significant advantages:

- they are non-invasive and can operate without direct contact with the target surface, eliminating problems related to heat transfer delays, sensor degradation, or contamination;
- they are stable and reliable at very high temperatures, where contact sensors may fail due to material limitations;
- their accuracy remains high also in environments up to and exceeding 1450 °C, making them particularly suitable for industrial process control, metallurgical furnaces, glass and ceramics production, and combustion system;
- their sensitivity increases with temperature, which means they perform better at high temperatures than at low ones.

On the other hand, pyrometers are relatively expensive instruments, and their accuracy can be affected by variations in surface emissivity, optical alignment, and interfering radiation from the surrounding environment. Despite these limitations, their ability to provide rapid, stable, and non-contact measurements makes them indispensable for monitoring and controlling thermal processes in the range of approximately −50 to 3000 °C (depending on the pyrometer type), especially in contexts where direct sensor placement is not feasible or would compromise measurement integrity. In conclusion, pyrometers represent a sophisticated class of temperature sensors that translate the radiative power emitted by a surface into an electrical signal. By leveraging the fundamental principles of thermal radiation described by Planck and Stefan-Boltzmann, they provide a robust and precise means of temperature measurement, particularly suited for high-temperature, non-contact, and industrial applications.

Each of these sensor types presents distinct advantages and limitations concerning sensitivity, dynamic range, response time, and durability. The optimal choice depends on the intended application, the environmental conditions, and the required measurement accuracy. In particular, in gas-sensing measurement chambers used for the characterization of chemoresistive sensors, temperature monitoring of the surrounding atmosphere is essential to ensure reliable and reproducible data. The most commonly employed temperature sensors in this context are platinum resistance thermometers (RTDs, typically Pt100 or Pt1000), due to their excellent stability, linear response, and chemical inertness, which make them ideal for long-term laboratory applications. Thermistors, usually of the NTC type, are also frequently used when higher sensitivity within a limited temperature range is required, providing rapid response and cost-effectiveness despite their non-linear behavior. Thermocouples (often of type K or T) are preferred in systems operating at higher temperatures or under harsh environmental conditions because of their robustness and wide operating range, though they offer lower precision. In more compact or automated setups, digital semiconductor-based temperature sensors are sometimes integrated for convenience, but their use remains limited to moderate temperatures.

9.3 Humidity Sensors

Humidity plays a crucial role in gas-sensing applications, as the amount of water vapor present in the surrounding atmosphere can significantly influence the response of chemoresistive sensors. The measurement of humidity quantifies the concentration of water vapor in a gas, which can either be a mixture, such as air, or a pure gas, such as nitrogen or argon. Different parameters are used to describe moisture content, including Relative Humidity (RH), a temperature-dependent, relative measurement, and Dew/Frost Point or Parts Per Million (PPM), which are absolute measurements independent of temperature. In Table 9.2 a list of the main parameters is shown.

Humidity sensors are generally classified into Relative Humidity (RH) sensors and Absolute Humidity (AH) sensors (or moisture sensors), with the former being the most widely employed due to their versatility and ease of integration. Humidity can be measured through traditional instruments such as dry and wet bulb hygrometers or dew point hygrometers, but modern research and industrial applications increasingly rely on electronic hygrometers, commonly referred to as humidity sensors, which offer improved precision, miniaturization, and real-time monitoring capabilities. Polymeric humidity sensors can be classified into various types, as illustrated in Fig. 9.8. However, in the context of chemoresistive sensor applications, the devices typically employed are electronic humidity sensors; therefore, the following discussion will focus specifically on this category.

Electronic humidity sensors operate based on the variation of electrical properties (typically capacitance or resistance) caused by changes in the water vapor content of the surrounding atmosphere. Among them, **capacitive humidity sensors** are the most employed due to their stability, repeatability, and wide measurement range. These devices consist of hygroscopic dielectric material, generally a polymer or plastic layer with a dielectric constant between 2 and 15, sandwiched between two conductive electrodes to form a small capacitor. In dry conditions, the capacitance

Table 9.2 Parameters used to describe moisture content

Parameter	Definition	Unit
Absolute humidity (AH)	Ratio of mass (vapor) to volume	g/m^3
Mass ratio	Ratio of mass (vapor) to mass (dry gas)	–
Relative humidity (RH)	Ratio of mass (vapor) to mass (saturated vapor)	%
Specific humidity	Ratio of mass (vapor) to total mass	%
Dew point	Temperature at which the water vapor in a gas condenses to water	°C
Frost point	Temperature at which the water vapor in a gas solidifies to ice	°C
Volume ratio	Ratio of partial pressure (vapor) to partial pressure (dry gas)	%
PPM by volume	Ratio of volume (vapor) $\times 10^6$ to volume (dry gas)	ppm_V
PPM by weight	Ratio of mass (vapor) $\times 10^6$ to mass (dry gas)	ppm_W

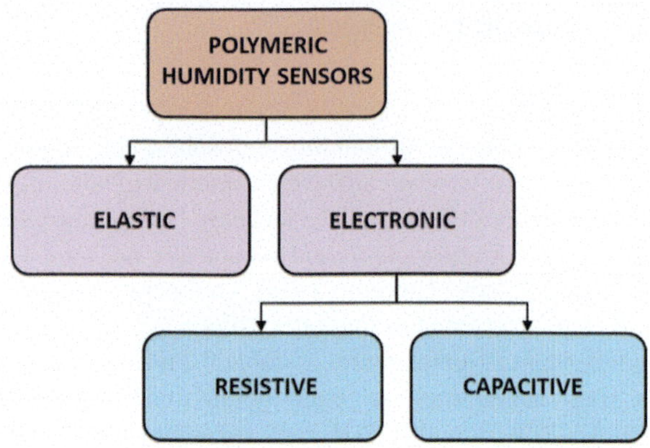

Fig. 9.8 Main polymeric humidity sensor types

is determined by the dielectric constant of the polymer and the sensor geometry; however, since water vapor has a much higher dielectric constant (approximately 80 at room temperature), its absorption into the dielectric leads to a measurable increase in capacitance. At equilibrium, the amount of moisture within the sensing layer depends on both the ambient temperature and the partial pressure of water vapor, establishing a direct relationship between the RH and the sensor's capacitance. Structurally, a typical capacitive humidity sensor is fabricated on an alumina substrate, with the bottom electrode composed of a noble metal such as gold or platinum. A humidity-sensitive polymer film, for example polyvinyl alcohol (PVA), is deposited over the electrode, and a semi-permeable top electrode (usually a thin gold layer) is added, allowing water vapor to diffuse into the sensing layer until equilibrium is reached. This configuration ensures both electrical conductivity and efficient vapor exchange, enabling accurate and reversible RH measurements. In Fig. 9.9 a capacitive humidity sensor is schematized.

Another important category includes **resistive humidity sensors**, which detect humidity through variations in the electrical resistance of the sensing film. In these devices, as schematized in Fig. 9.10, comb-shaped electrodes made of materials such as gold or ruthenium oxide are printed on a substrate and covered with a polymeric

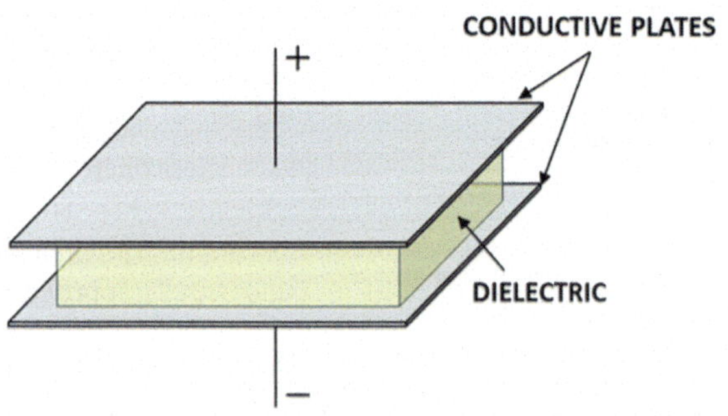

Fig. 9.9 Basic scheme of a capacitive humidity sensor

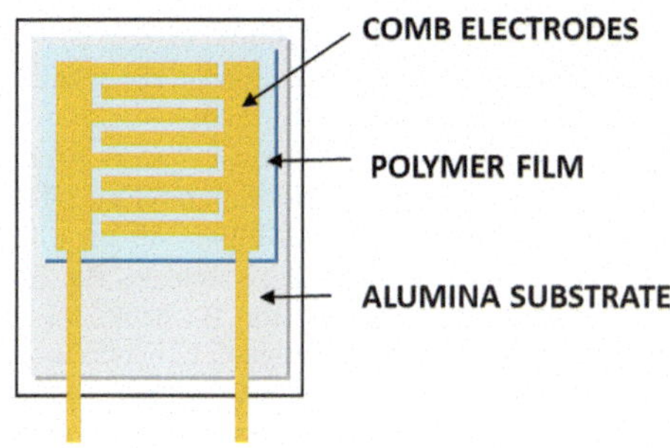

Fig. 9.10 Schematization of a resistive humidity sensor

layer containing mobile ions. As water molecules are absorbed, the ionic conductivity of the polymer increases, leading to a significant decrease in resistance (or impedance). This behavior provides a simple yet effective mechanism for humidity detection, particularly suitable for low-cost and compact sensing systems.

Overall, capacitive and resistive humidity sensors represent the two dominant technologies in modern humidity measurement, combining high sensitivity, miniaturization potential, and compatibility with both laboratory instrumentation and integrated environmental monitoring systems.

References

1. Incropera FP, DeWitt DP, Bergman TL, Lavine AS (2007) Fundamentals of heat and mass transfer. Wiley, Hoboken
2. Haynes WM (2014) CRC handbook of chemistry and physics. CRC Press, Boca Raton
3. ASM International (1990) ASM handbook, vol 2. Properties and selection: nonferrous alloys and special-purpose materials. ASM International, Materials Park
4. Barron RF (1985) Cryogenic systems, 2nd edn. Oxford University Press, Oxford
5. Jha CM (ed) (2015) Thermal sensors: principles and applications for semiconductor industries. Springer, New York
6. Fraden J (2010) Handbook of modern sensors: physics, designs, and applications, 4th edn. Springer, New York
7. Giberti A, Benetti M, Carotta MC, Guidi V, Malagù C, Martinelli G (2008) Heat exchange and temperature calculation in thick-film semiconductor gas sensor systems. Sens Actuators B 130:277–280
8. Giberti A, Carotta MC, Guidi V, Malagù C, Martinelli G, Milano L (2009) Influence of ambient temperature on electronic conduction in thick-film gas sensors. Sens Actuators B Chem 137(1):111–114
9. Korotcenkov G (2020) Handbook of humidity measurement: methods, materials and technologies. CRC Press, Boca Raton. ISBN 978-1-138-29787-6
10. Chani MT, Asiri AM, Khan SB (eds) (2023) Humidity sensors: types and applications. IntechOpen, London. ISBN 978-1-83968-565-1
11. Gherardi S, Astolfi M, Gaiardo A, Malagù C, Rispoli G, Vincenzi D, Zonta G (2024) Investigating the temperature-dependent kinetics in humidity-resilient tin–titanium-based metal oxide gas sensors. Chemosensors 12(8):151
12. Gherardi S, Zonta G, Astolfi M, Malagù C (2021) Humidity effects on SnO2 and (SnTiNb)O2 sensors response to CO and two-dimensional calibration treatment. Mater Sci Eng B 265:115013

13. Giberti A, Carotta MC, Guidi V, Malagù C, Martinelli G, Piga M, Vendemiati B (2004) Monitoring of ethylene for agro-alimentary applications and compensation of humidity effects. Sens Actuators B 103:272–276
14. Spagnoli E, Fabbri B, Gaiardo A, Valt M, Ardit M, Krik S, Cruciani G, Della Ciana M, Vanzetti L, Vola G, Di Benedetto F, Migliori A (2022) Design of a metal-oxide solid solution for selective detection of ethanol with marginal influence by humidity. Sens Actuators B Chem 370:132426
15. Traversa E (1995) Ceramic sensors for humidity detection: state-of-the-art and future developments. Sens Actuators B Chem 23(2–3):135–156
16. Ortega PP, Gherardi S, Spagnoli E, Fabbri B, Astolfi M, Zonta G, Landini N, Malagù C, Aldao CM, Ponce MA, Simoes AZ, Longo E (2024) Nanostructured Eu-doped ceria for humidity sensing: a morphological perspective. Appl Surf Sci 661:160047

Chapter 10
Measurement Setups and Applications

If you only read the books that everyone else is reading,
you can only think what everyone else is thinking.
Haruki Murakami

The laboratory testing phase represents a crucial stage in the development of chemoresistive sensors, as it enables a thorough characterization of their electrical and functional properties. During this phase, the sensor responses are systematically analyzed under exposure to different target gases and varying concentrations, while simultaneously determining the optimal operating temperature of the integrated heater. Environmental parameters, such as humidity, are also adjusted to reproduce realistic conditions consistent with the intended application and the specific gas to be detected. The analysis of the collected data provides comprehensive insights into the electrical behavior of the sensor, allowing for the optimization of operating parameters to enhance both sensitivity and selectivity toward specific analytes. This stage is therefore fundamental for assessing the sensor's performance, efficiency, and discrimination capability. Moreover, in addition to the laboratory setup, there is also the possibility of inserting the sensors into specific portable measuring units, suitable for different applications, each composed of specific components for reading and analyzing the signal. In particular, chemical sensors currently find application across several major markets, including toxic and combustible gas sensing, automotive technologies, energy conservation, medical and healthcare fields, as well as agrifood, industrial, environmental, and aerospace applications. Each portable device must be adapted to the specific industrial need. In this Chapter, there is a basic description of both the laboratory setup and a typical portable device for gas sensing applications. Drawing on an integrated review of the sources cited here, this Chapter includes the authors' original contributions, with selected references cited explicitly where relevant [1–17].

© The Author(s), under exclusive license to Springer Nature Switzerland AG 2026

C. Malagù and G. Zonta, *Gas Sensors*, https://doi.org/10.1007/978-3-032-21614-4_10

10.1 Laboratory Setup

A laboratory experimental apparatus is designed to allow precise regulation and monitoring of several key parameters, including:

- the temperature inside the test chamber;
- the sensor's operating temperature;
- the gas concentration within the chamber;
- the water vapor partial pressure.

The overall setup consists of multiple parts, each performing a specific function. These include a pneumatic system for gas handling and mixing, an electronic control system for temperature and signal acquisition management, and a measurement chamber where the sensors are exposed to the gas mixture under controlled conditions. Let's analyze the main parts.

The **pneumatic system** is responsible for delivering a controlled gas mixture to the test chamber, where the sensors are housed. The gases required for the measurements are supplied from cylinders (located outside the laboratory for security reasons) and conveyed through dedicated stainless-steel lines. The system includes a carrier gas flow, composed of synthetic air (20% oxygen and 80% nitrogen), along with dedicated lines for the various gas mixtures to be analyzed. The gas flow is regulated by Mass Flow Controllers (MFCs), each equipped in series with an electromechanical valve. This configuration enables precise control of the individual gas flows, allowing the preparation of mixtures with the desired concentrations. To simulate humid air (wet conditions), to which the sensor may be exposed during sample analysis, the dry air stream is passed through a bubbler, a device containing a porous septum immersed in deionized water. As the air passes through the water, it becomes saturated with moisture, and the resulting humid mixture is then conveyed into the measurement chamber through four lateral inlets, ensuring a uniform gas distribution. After interacting with the sensors, the gas is discharged through an exhaust tube located on the lid of the testing chamber, which is connected to a fume extraction system. A schematic representation of the pneumatic system is shown in Fig. 10.1.

The **measurement chamber** (Fig. 10.2) is placed inside a thermostatic enclosure that allows the desired temperature to be set and maintained throughout the measurements, thereby stabilizing the ambient temperature. Internally, the measurement chamber accommodates up to eight gas sensor housings and is equipped with humidity and temperature monitoring systems.

As already explained in Chap. 9, accurate knowledge and control of these two parameters are crucial, as environmental conditions can significantly influence the sensor responses and, consequently, the overall reliability of the measurements.

The **electronic system** is responsible for regulating and monitoring all stages involved in the sensor electrical characterization. Due to the complexity associated with managing these components, the use of a computer as an interface for communication with the various devices is essential. Communication is established according to the IEEE standard through a dedicated software platform. The computer

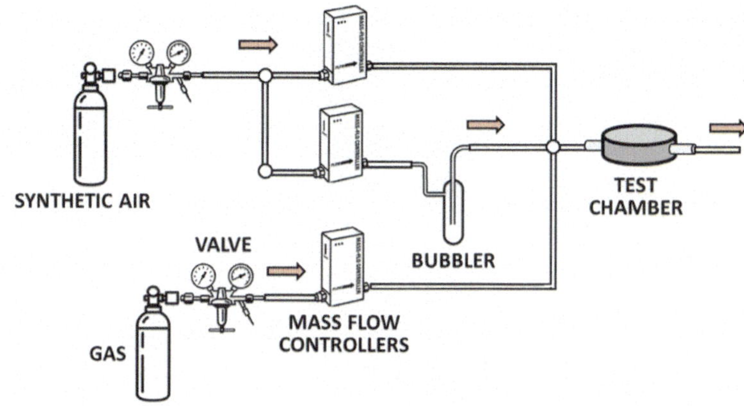

Fig. 10.1 The first gas line supplies dry air, while the second line delivers humidified air generated by passing the carrier gas through a bubbler. The third line is dedicated to the introduction of the target analyte. Mass flow controllers (MFCs) are employed to precisely regulate and mix the individual gas flows, thereby controlling the total concentration of the target gas reaching the sensors

Fig. 10.2 Test chamber with eight sensor allocations, a temperature sensor and a humidity sensor hosted in the center and the charge resistances in the circuit on the left

directly interacts with the programmable power supplies and the digital multimeter, while it communicates indirectly with the measurement chamber via a dedicated electronic board, using the aforementioned devices as intermediaries. A schematic representation of the electronic unit is shown in Fig. 10.3.

The power supplies deliver the voltage V_H, which is required to heat the resistive element of the sensor's heater. The digital multimeter, on the other hand, records the voltage signals associated with the eight chemoresistive sensors, as well as with the humidity and temperature sensors. All these parameters can be visualized in real

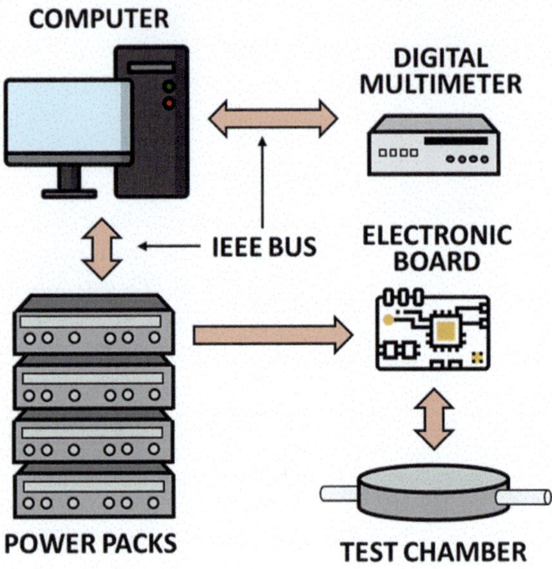

Fig. 10.3 Schematic representation of the experimental setup main components

time on the computer screen through a custom-made application, whose interface is schematized in Fig. 10.4.

The temperature of the sensing film must be carefully controlled to ensure the proper operation of the sensor. Regulation is achieved through the integrated heater, which raises the temperature of the sensitive layer by Joule heating. Exploiting the physical principle that the electrical resistance of a conductor varies with temperature, it is possible to regulate the sensor temperature according to the following relation:

$$R(T) = R_0\left(1 + \alpha T + \beta T^2\right),$$

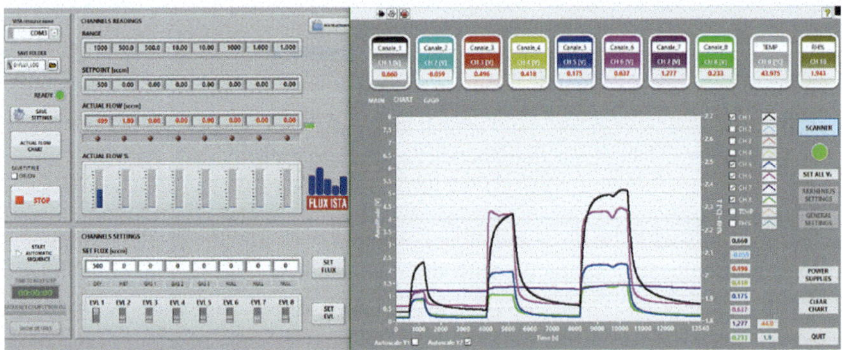

Fig. 10.4 Measurement software interface: the left panel corresponds to the gas flow control, while the right panel the electrical characterization

where $\alpha = 3.380 \times 10^{-3}$ °C^{-1} and $\beta = -6.826 \times 10^{-7}$ °C^{-2} are the thermal coefficients of resistivity, experimentally determined for the material. The resistance at 0 °C, denoted R_0, can be obtained by measuring the resistance at a known temperature environment. Rearranging the previous relation gives:

$$R_0 = \frac{R(T)}{1 + \alpha T + \beta T^2}.$$

Then, by combining the coefficients α and β with the reference value R_0, it becomes possible to determine the resistance corresponding to a desired temperature, thereby associating each operating temperature with an appropriate resistance value. In practice, a voltage V_H is applied to the heater so that its resistance matches the one required to reach the target temperature. In other words, V_H is adjusted such that the heater resistance R assumes the desired value, which is directly related to the sensor temperature T. The sensor response is investigated by measuring its conductance variations. At fixed time intervals (generally 5 s), the voltage V_{out} of the electronic circuit containing the sensor is recorded using the digital multimeter. The circuit is configured as an inverting operational amplifier, which amplifies the sensor signal and makes conductance variations clearly observable. The circuit schematic is shown in Fig. 10.5.

In the circuit, V_{in} represents the bias voltage (set to −5 V), R_f is the known feedback resistance (selected by the operator to modulate the sensor baseline level and to avoid signal saturation) and R_S corresponds to the resistance of the sensing film. The transfer function of the circuit is expressed as:

$$V_{out} = -\frac{R_f}{R_S} V_{in},$$

from which the sensor resistance R_S can be derived: $R_S = -\frac{V_{in}}{V_{out}} R_f$.

Since conductance is defined as the reciprocal of resistance, the sensor conductance G can be expressed as:

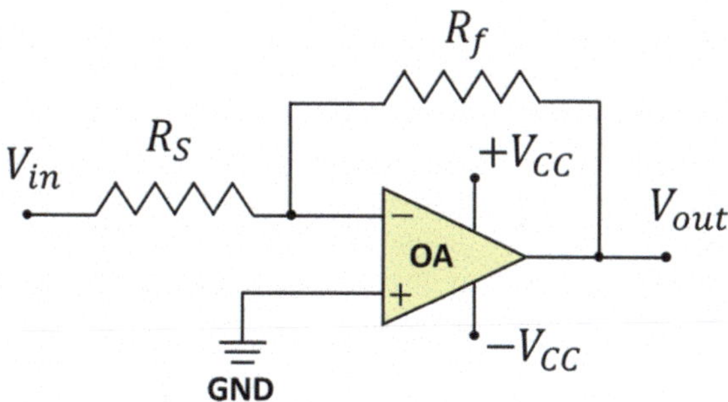

Fig. 10.5 Inverting operational amplifier: the circuit in which the sensor is inserted

$$G = -\frac{1}{R_f}\frac{V_{out}}{V_{in}}.$$

Once the conductance has been determined, a normalization procedure is applied to derive the sensor response S (which is an dimensionless quantity), by dividing the conductance measured in the presence of the target gas ($G_{gas} = - R_f^{-1}V_{gas}V_{in}^{-1}$) by that measured in synthetic air ($G_{air} = - R_f^{-1}V_{air}V_{in}^{-1}$), where V_{gas} and V_{air} denote the voltages measured in gas and air, respectively, obtaining:

$$S = \frac{G_{gas}}{G_{air}} = \frac{V_{gas}}{V_{air}}.$$

Sometimes it is useful to normalize the response to zero, to ensure a null signal in the absence of the target gas. This approach yields the relative variation of conductance with respect to the baseline. The overall sensor response as a function of gas concentration is therefore given by:

$$S_{\text{normalized}} = \frac{G_{gas}}{G_{air}} - 1 = \frac{V_{gas}}{V_{air}} - 1.$$

10.2 Portable Setups

Chemoresistive sensors have a wide range of application domains, including industrial, environmental, automotive, biomedical, space, homeland security, and agri-food sectors. Each of these fields requires specific sensing materials (employed either individually or in arrays comprising several elements) along with sensor calibrations tailored to the specific purpose of the application. These calibrations are often complex and rely on highly customized software platforms, which may vary substantially in their level of sophistication, ranging from simple signal acquisition and processing routines to advanced recognition algorithms based on machine learning or artificial intelligence techniques. The corresponding experimental setups also differ greatly in their configuration (whether fixed or portable) in terms of design, casing, pneumatic system, geometry of the measurement chambers, and overall size and materials. Such diversity arises both from the differences in the intended applications (e.g., the safety standards required for use in medical environments or in close contact with humans) and from the operating environments (e.g., outdoor conditions subject to meteorological variability versus controlled indoor settings). Therefore, although all systems are fundamentally based on the same core laboratory components, i.e., a computer (either integrated or external, possibly equipped with an embedded display), dedicated signal acquisition and analysis software, a pneumatic system, an electronic apparatus, and the sensing unit (including the measurement chambers and sensors), each setup may differ considerably from the others in its structural and functional implementation. For these reasons, the development of each measuring device requires both a feasibility study, essential for determining the appropriate materials, the number of sensors, and the overall design of the instrument, and a more extensive investigation under real operating conditions. As an illustrative example, this Section describes one of the devices developed at the Sensor Laboratory of the Department of Physics and Earth Sciences of the University of Ferrara (UNIFE). The system, for which both national and international patents have been granted, was designed for colorectal cancer screening through the analysis of human fecal volatile emissions.

Research leading to the development of the device began in 2013, with the aim of detecting colorectal cancer (CRC) through its volatile biomarkers. CRC is among the most prevalent malignancies worldwide and the second leading cause of cancer-related deaths after lung cancer, however, when diagnosed early, CRC is highly curable (approximately 90% survival at stage I), highlighting the crucial role of preventive screening. A variety of chemoresistive materials were arranged into sensor arrays to detect volatile organic compounds (VOCs) reported as CRC biomarkers in exhaled breath under the hypothesis that similar VOCs could also be present in intestinal gases. To reproduce such an environment, an artificial intestine was developed in the laboratory, consisting of a hermetically sealed sensor chamber connected via Teflon tubing to standard gas cylinders and mass flow controllers, operated by dedicated PC software. Sensor responses were recorded both for individual CRC biomarkers and for mixtures containing intestinal interferents at typical physiological concentrations. Previous studies using canine olfaction, gas chromatography, and commercial electronic noses had already shown that CRC alters fecal volatile emissions. Building on these findings, SL designed a device for non-invasive CRC screening through fecal exhalation analysis. The resulting device (shown in Fig. 10.6) comprised a microfluidic sampling system, dedicated electronics, and a sensing unit with an array of metal-oxide sensors [16].

Currently, the fecal immunochemical test (FIT) is the most common CRC screening method, performed on individuals but it suffers from a high false-positive rate to CRC. A preliminary study was conducted in collaboration with the S. Anna University Hospital of Ferrara, involving fecal samples from CRC patients and healthy volunteers and, following these promising results, the device underwent a three-year clinical validation trial (2016-2019), approved by the Ethics Committee and involving UNIFE, S. Anna Hospital, the Department of Public Health, and SCENT S.r.l. Around 1000 individuals with positive FIT results were enrolled prior to colonoscopy. Double-blind tests demonstrated specificity and sensitivity of 82.4% and 84.1%, achieved through Support Vector Machine (SVM) analysis, increased device reproducibility and suitability for commercialization. The device is robust, portable, and user-friendly and the more optimized version achieved CE certification in 2023. The measurement protocol is standardized and semi-automated: patients collect fecal samples in dedicated containers, store them at -18 °C, and introduce them into the device after thawing. A single touchscreen command initiates the analysis, which lasts about ten minutes. The result (positive or negative) is displayed

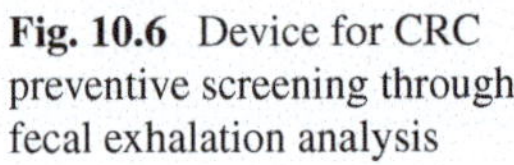
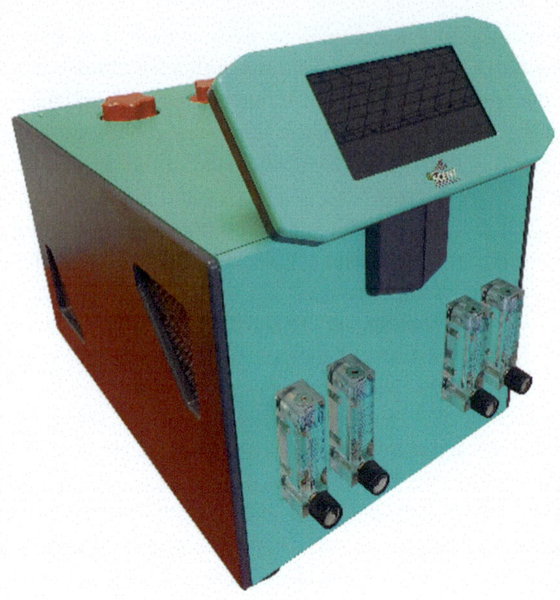

Fig. 10.6 Device for CRC preventive screening through fecal exhalation analysis

automatically through an algorithm developed by the SL team, which analyzes the dynamic response curves of the sensors with high precision.

Another portable device developed at the SL is a patented portable prototype for non-invasive cancer screening and monitoring via blood headspace analysis, and for fundamental oncological research through VOCs emitted by cell cultures. Developed in collaboration with the Department of Neuroscience and Rehabilitation at UNIFE, it retains the core architecture of the one for colorectal cancer screening, comprising a pneumatic unit, an electronic acquisition module, and a chemoresistive sensing array. Adaptations in the sensing core, using tailored materials, optimize performance for specific applications. The device detects VOCs released by highly vascularized tumors, enabling correlation between sensor responses and tumor presence, vascularization, and disease progression. Non-vascularized tumors, lacking bloodstream contact, do not produce detectable signals, highlighting the specificity of the approach. The device has been applied to longitudinal analyses of blood from CRC patients across pre- and post-surgical stages, successfully discriminating early post-operative samples from long-term follow-up. Individual sensors varied in response amplitude and sensitivity, yet the combined array provided reliable monitoring of patient status. In summary, this device offers a compact, non-invasive, and versatile platform for VOC-based blood analysis, supporting tumor detection, therapeutic assessment, and post-treatment follow-up. Its performance underscores its potential as a complementary tool to conventional diagnostics within personalized and minimally invasive oncology. In Fig. 10.7 the prototype [17].

In conclusion, the versatility of current chemoresistive sensing technology enables the adaptation of portable devices to a wide range of applications across medical, environmental, industrial, space, and agri-food sectors. At present, multiple research

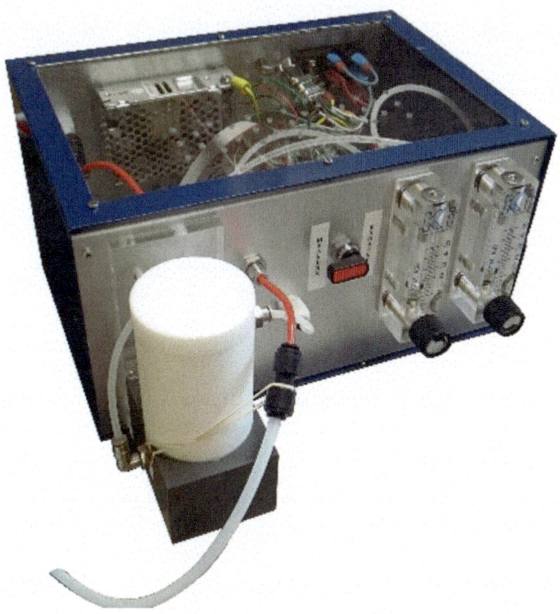

Fig. 10.7 First prototype for tumor monitoring and basic oncologic research through blood and cell cultures exhalation analysis

lines exploring these diverse applications are actively pursued at the SL, reflecting the broad potential and ongoing innovation of this field.

References

1. Astolfi M, Zonta G, Malagù C, Anania G, Rispoli G (2025) PMOX nanosensors to detect colorectal cancer relapses from patient's blood at three years follow-up, and gender correlation. Biosensors (Basel) 15(1):56
2. Astolfi M, Rispoli G, Anania G, Zonta G, Malagù C (2023) Chemoresistive nanosensors employed to detect blood tumor markers in patients affected by colorectal cancer in a one-year follow up. Cancer 15(6):1797
3. Astolfi M, Rispoli G, Benedusi M, Zonta G, Landini N, Valacchi G, Malagù C (2022) Chemoresistive sensors for cellular type discrimination based on their exhalations. Nano 12(7):1111
4. Zonta G, Malagù C, Gherardi S, Giberti A, Pezzoli A, De Togni A, Palmonari C (2020) Clinical validation results of an innovative non-invasive device for colorectal cancer preventive screening through fecal exhalation analysis. Cancer 12(6):1471
5. Astolfi M, Rispoli G, Anania G, Artioli E, Nevoso V, Zonta G, Malagù C (2021) Tin, titanium, tantalum, vanadium and niobium oxide based sensors to detect colorectal cancer exhalations in blood samples. Molecules 26(2):466
6. Landini N, Anania G, Astolfi M, Fabbri B, Guidi V, Rispoli G, Valt M, Zonta G, Malagù C (2020) Nanostructured chemoresistive sensors for oncological screening and tumor markers tracking: single sensor approach applications on human blood and cell samples. Sensors 20(5):141. https://doi.org/10.3390/s2005141
7. Astolfi M, Rispoli G, Anania G, Nevoso V, Artioli E, Landini N, Benedusi M, Melloni E, Secchiero P, Tisato V, Zonta G, Malagù C (2020) Colorectal cancer study with nanostructured

sensors: tumor marker screening of patient biopsies. Nano 10(4):606. https://doi.org/10.3390/nano10040606

8. Zonta G, Anania G, Feo C, Gaiardo A, Gherardi S, Giberti A, Guidi V, Landini N, Palmonari C, Ricci L, de Togni A, Malagù C (2018) Use of gas sensors and FOBT for the early detection of colorectal cancer. Sens Actuators B Chem 262:884–891
9. Zonta G, Anania G, Fabbri B, Gaiardo A, Giberti A, Landini N, Malagù C, Scagliarini L, Guidi V (2017) Preventive screening of colorectal cancer with a device based on chemoresistive sensors. Sens Actuators B Chem 238:1098–1101
10. Zonta G, Anania G, Fabbri B, Gaiardo A, Gherardi S, Giberti A, Guidi V, Landini N, Malagù C (2015) Detection of colorectal cancer biomarkers in the presence of interfering gases. Sens Actuators B Chem 218:289–295
11. Zonta G, Anania G, Astolfi M, Feo C, Gaiardo A, Gherardi S, Giberti A, Guidi V, Landini N, Palmonari C, de Togni A, Malagù C (2019) Chemoresistive sensors for colorectal cancer preventive screening through fecal odor: Double-blind approach. Sens Actuators B Chem
12. Landini N, Anania G, Fabbri B, Gaiardo A, Gherardi S, Guidi V, Rispoli G, Scagliarini L, Zonta G, Malagù C (2018) Neoplasms and metastasis detection in human blood exhalations with a device composed by nanostructured sensors. Sens Actuators B Chem 271:203–214
13. Gaiardo A, Zonta G, Gherardi S, Malagù C, Fabbri B, Landini N, Vanzetti L, Landini N, Casotti D, Cruciani G, Della Ciana M, Guidi V (2020) Nanostructured SMFEO3 gas sensors: investigation of the gas sensing performance reproducibility for colorectal cancer screening. Sensors 20(20):5910
14. Guidi V, Carotta MC, Fabbri B, Gherardi S, Giberti A, Malagù C (2012) Array of sensors for detection of gaseous malodors in organic decomposition products. Sens Actuators B Chem 174:349–354
15. Malagù C, Fabbri B, Gherardi S, Giberti A, Guidi V, Landini N, Zonta G (2014) Chemoresistive gas sensors for the detection of colorectal cancer biomarkers. Sensors 14:18982–18992
16. Zonta G, Gherardi S, Giberti A, Landini N, Gaiardo A (2014) Device for preliminary screening of colorectal adenomas. Italian national patent RM2014A000595; European patent EP3210013 (validated in Germany and the United Kingdom), extension granted in 2018
17. Malagù C, Gherardi S, Zonta G, Landini N, Giberti A, Fabbri B, Gaiardo A, Anania G, Rispoli G, Scagliarini L (2015) Combination of nanoparticulate semiconductor materials for distinguishing normal cells from tumor cells Italian national patent IT102015000057717

The manufacturer's authorised representative in the EU is Springer Nature Customer Service Centre GmbH, Europaplatz 3, 69115 Heidelberg, Germany. If you have any concerns regarding our products, please contact ProductSafety@springernature.com

Printed and bound by CPI Group (UK) Ltd, Croydon, CR0 4YY
07/07/2026
02160909-0001